Postmodern

Postmodern Geography

Theory and Praxis

Edited by Claudio Minca
University of Venice

BLACKWELL *Publishers*

Copyright © Blackwell Publishers Ltd 2001
Editorial matter and arrangement copyright © Claudio Minca 2001

The moral right of Claudio Minca to be identified as author of the Editorial Material
has been asserted in accordance with the Copyright, Designs and Patents Act 1988.

First published 2001

2 4 6 8 10 9 7 5 3 1

Blackwell Publishers Ltd
108 Cowley Road
Oxford OX4 1JF
UK

Blackwell Publishers Inc.
350 Main Street
Malden, Massachusetts 02148
USA

British Library Cataloguing in Publication Data

A CIP catalogue record for this book is available from the British Library.

Library of Congress Cataloging-in-Publication Data

Postmodern geography: theory and praxis / edited by Claudio Minca.
     p. cm.
Includes bibliographical references and index.
   ISBN 0–631–22559–5 (hbk. : acid-free paper) — ISBN 0–631–22560–9
(pbk. : acid-free paper)
   1. Urban geography.   2. Human geography—Philosophy.   I. Minca,
Claudio.
   GF 125 .P67 2001
   910′.9173′2—dc21

                                                              00–011045

Typeset in 10½ on 12 pt Sabon
by Best-set Typesetter Ltd., Hong Kong
Printed in Great Britain by TJ International, Padstow, Cornwall

This book is printed on acid-free paper.

Contents

Figures

Contributors

Denis Cosgrove is Professor of Geography at Royal Holloway College, University of London. He has written on questions of landscape and representation and is the editor of *The Iconography of Landscape* (with S. Daniels, 1988) and *Mappings* (1999). His books include *Social Formation and Symbolic Landscape* (1984), *The Palladian Landscape* (1993), and *Apollo's Geography: Global Images and Meanings in Western Culture* (2000).

Michael Dear is Professor of Geography and the Director of the Southern California Studies Center at the University of Southern California. He is the founding editor of *Society and Space* (*Environment and Planning D*) and the author of numerous articles on postmodernism in geography. His most recent books include *Malign Neglect: Homelessness in an American City* (with J. Wolch, 1996) and *The Postmodern Urban Condition* (2000).

Giuseppe Dematteis is Professor of Urban Geography at the University of Torino. He has written extensively on the metaphorical meaning of geographical representations and urban restructuring in Italy and is the author of *Le metafore della Terra* (1985), *Progetto implicito* (1995) and the editor (with V. Guarrasi) of *Urban Networks* (1995).

Franco Farinelli is Professor of Geography at the University of Bologna and the author of an extensive body of work on the history of geography and cartography and the theory of geographical representations. His books include *Pour une théorie générale de la géographie* (1989), *I segni del mondo* (1992) and he is the editor (with G. Olsson and D. Reichert) of *Limits of Representation* (1993).

Steven Flusty is a doctoral student in the Department of Geography at the University of Southern California. He has written on questions of postmodern urbanism and is the author of *Building Paranoia: The Proliferation of Interdictory Space and the Erosion of Spatial Justice* (1994).

Vincenzo Guarrasi is Professor of Geography at the University of Palermo. He has published widely on cultural geographic issues and the production of urban space inspired by the work of Henri Lefebvre; is the author of *La condizione marginale* (1978), *La produzione dello spazio urbano* (1981) and the editor (with G. Dematteis) of *Urban Networks* (1995).

Cindy Katz is Professor of Geography at the Graduate School of the City University of New York and is the author of numerous articles on the production of space, place, and nature in everyday life. She is the co-editor (with J. Monk) of *Full Circles: Geographies of Women over the Life Course* (1994).

Luciana L. Martins is a postdoctoral research fellow at Royal Holloway, University of London. She is the author of *O Rotas do Saber: O Rio de Janeiro dos Viajantes Britanicos, 1800–1850* (2000). She is currently working on an AHRB-funded project on "Knowing the Tropics: British Visions of the Tropical World, 1750–1850."

Claudio Minca is Associate Professor of Political and Economic Geography at the University of Venezia. He has written widely on geographical representations and postmodernism in geography and is the author of *Spazi effimeri* (1996) and the editor of a forthcoming collection entitled *Introduzione alla geografia postmoderna*.

Don Mitchell is Associate Professor of Geography at Syracuse University. He has written extensively on the control of public space and is the author of *The Lie of the Land: Migrant Workers and the California Landscape* (1996) as well as of *Cultural Geography: A Critical Introduction* (2000).

Gunnar Olsson is Professor of Geography at the University of Uppsala. He is the author of *Birds in Egg/Eggs in Bird* (1980), *Lines of Power/Limits of Language* (1991), and of numerous articles and essays on questions of representation and language in geography. He is also the editor (with F. Farinelli and D. Reichert) of *Limits of Representation* (1993).

Neil Smith is Professor of Geography at Rutgers University. He is the author of countless articles on political economy, geographical theory, and historical studies of space and nature, as well as *Uneven Development: Nature, Capital and the Production of Space* (1984) and *The New Urban Frontier: Gentrification and the Revanchist City* (1996), and is co-editor of *Gentrification of the City* (1986) and *Geography and Empire* (1994).

Edward W. Soja is Professor of Urban and Regional Planning at the Graduate School of Architecture and Urban Planning at the University of California, Los Angeles. He has written widely about the place of space in contemporary social theory, notions of spatiality, and theorizations of postmodernity. He is the author of *Postmodern Geographies* (1989), *Thirdspace* (1996), and *Postmetropolis: Critical Studies of Cities and Regions* (2000).

Acknowledgments

Despite the fact that it is my name as editor which appears on the cover, this volume like any other is the result of the efforts of many people, all of whom have contributed to its creation. The idea for this edited collection was born during a conference held in Venezia in June of 1999, whose theme – 'Postmodern Geographical Praxis' – was to inspire the direction of the present work. The meeting, which first brought together the scholars present within the pages of this volume, was made entirely possible by the Fondazione ENI Enrico Mattei here in Venezia, to whom I had presented a proposal for the encounter and who graciously agreed to host the conference. My thanks thus go, above all, to the Fondazione and to its staff for the impeccable organization of this event which opened the way to the reflection within the pages to follow.

I owe a particular debt to two geographers from across the ocean, Michael Dear and Edward Soja, who not only inspired my own voyage into postmodern geography but also provided me with their support and guidance during the various phases of this particular project. Thank you for your invaluable help in all the critical moments that accompanied this initiative. Gabriele Zanetto, friend, colleague, and the founding father of our 'Venetian School,' gave his unwavering support to this initiative from its earliest stirrings, just as he has always championed even the most foolhardy of my initiatives. A special thanks, of course, also goes to all of the authors who agreed to contribute to this volume and, in doing so, allowed me to realize a project that I had long dreamt of putting in print.

Sarah Falkus and Katherine Warren, my editors at Blackwell, steered this volume through its various stages with patience and care and helped remedy the variety of technical and organizational mishaps that came with my editorial tasks. Joanna Pyke and Cameron Laux guided me through all the practicalities of putting the book in print. Rachele Borghi generously contributed her time and skill (and her new laptop . . .) to the editing of the pieces. I cannot thank her enough for all the hours of work she put into the book, but also for her ever-present (and contagious) enthusiasm.

My biggest thanks go, however, to someone without whom this book would not have been possible at all. Luiza Bialasiewicz not only struggled with my own English but also helped with the translations of the other Italian contributions; her ideas, leads, and criticisms have been decisive in giving shape to this volume since its inception. Luiza accompanied and prodded me through the hard months of organizational work, shared the seemingly endless Sundays spent in front of the computer, and, most of all, always passionately believed in this project and its political significance. Un grazie di cuore Luiza.

Some argue that every book should be dedicated to someone; dedicating to this person or persons the efforts that have brought the work into being. Others are of the opinion, however, that dedicating a book to one's partner, family, or friends only brings bad luck – I'm not quite sure if to the book itself or to the unfortunate soul to whom it is dedicated. If I must dedicate this volume to someone, I would like to dedicate it to all those who have had the courage to try to free themselves of the strangleholds of modernity. I hope that this book can represent a sign that you are not alone.

The publishers apologize for any errors or omissions in copyright permission, and would be grateful to be notified of any corrections that should be incorporated into the text.

Prelude

Claudio Minca

The idea for this collection arose during a conference on postmodern geography that I had organized in Venice in 1999. It was that very meeting which first brought together the contributors to this volume and it was also then that many of the questions about the place and role of postmodern reflection within our discipline tackled within the pages of this book were first articulated. Nevertheless, the idea of publishing yet another edited collection on postmodern geography was not what I had in mind: first, because excellent works of this sort already exist, and secondly, because the task of recapitulating the past fifteen years of a long line of reflection within the Anglo-American academy certainly did not pertain to an 'external' observer such as myself.

The intentions behind this book were quite other. I wanted to pose a series of questions to some of the scholars shaping today's theoretical debates in geography on an issue that I had myself long held at heart: that is, what were the implications of the postmodern turn for geographical praxis? Or, what praxis for a postmodern geography? Running down the list of authors, the reader will note that I turned for answers both to some of today's leading geographers within the Anglo-American world, but also to key figures within my own, Italian, geographical tradition, for here the postmodern debate had developed in an entirely diverse fashion, and had been afforded an entirely diverse set of meanings, as will surely become apparent within the chapters written by Italian authors.

Why the need to problematize the meanings of a possible postmodern praxis in geography? Above all, because the postmodern debate within our discipline does not seem to have successfully resolved the key dilemma that has accompanied it since its inception: does the postmodern turn merely signal a new and revolutionary way of reading modernity (as many of the contributors to this volume seem to argue) – or do veritable postmodern spaces exist, are we now confronted with new and revolutionary geographies which follow coordinates that are radically different from those of modernity and thus necessitate new analytical tools, new understandings capable of grasping their dynamics?

We need to question ourselves about the possible meanings of a postmodern praxis in geography for another crucial reason as well, however; because, as we well know, the crisis of (geographical) representation has brought with it an accordant crisis of all 'projects' – that is, a growing difficulty if not impossibility of translating emergent theoretical reflection into operative suggestions able to dialogue with the complexity of the world. But can we attribute a strategic role to the emergent (postmodern?) geographical praxis beyond its deconstructionist exposure of the 'tricks' of normative, positivist knowledge? Can we still identify a political-projectual dimension to geography after the postmodern turn? And if (following the greatly overused saying) 'geography is what geographers do,' then we find ourselves back at our original dilemma: when we talk about postmodern geography, are we talking about new, purportedly postmodern readings of the (modern) world – or about new, revolutionary postmodern spaces for which we need new, revolutionary geographies? Thus, is a postmodern geographical praxis simply a *geographical reading* of the transformations within our societies, cultures, economies – a reading, in other words, of our new 'condition'? Or does any such praxis also actively *produce* such transformations (if we believe that all geographical readings of the world also necessarily transform this latter in practice)? The question is a crucial one, for it forces us to reconsider how the postmodern debate has fundamentally undermined what we imagine to be geography's and geographers' 'political' role.

I'd like to believe that the emergent challenges to modern ways of conceiving the world cannot but give rise to new geographical ways of knowing; new ways of knowing that eliminate, once and for all, the pernicious distinction between the theoretical and empirical moments, between theory and practice; new ways of knowing which, above all, are capable of imagining a new political-strategic role for geographical knowledge.

I thus offer the following disclaimer for the Anglo-American reader: this volume is *not* a review of state-of-the-art postmodern debate in geography, nor is it a continuation or recapping of multiple previous such attempts. Certainly, many of the chapters in this book draw upon this well-inscribed tradition – but many also do not, reminding us that geographers' reflections on the crisis of modernity do not necessarily travel a singular path but are, rather, the fruit of distinct cultural and academic contexts. This introduction intentionally lacks, in fact, an exhaustive literature review on postmodern geography, for any such review, again, would be partial at best, considering that the reflections presented within the pages of this volume emerge from diverse geographical traditions.

And if there is an explicit aim to the book, it is certainly that of presenting a series of widely diverse perspectives on the meanings and attributes of geographical praxis; perspectives which vary considerably both in the theoretical approaches adopted, but also in the distinct personal histories and positionings of the authors themselves – and thus necessarily in the 'political' objectives advanced. Like all edited collections of this sort, this volume too has both its advantages and disadvantages in this sense. On the one hand, all authors had to render explicit, in no uncertain terms, their position on the topic, articulating their distinct understandings of – and suggestions for – a postmodern geographical praxis. And this is, I believe, one of the strong points of the book. On the other hand, however, such diversity could make the whole appear fragmented at times. But this, again, is inevitable in any collection – and particularly one such as this which, unlike many other edited volumes, intentionally *did not* specify a unity of approach or a determinate theoretical framework.

Yet among all of these 'fragments' of reflection we *can* identify not only points of contact but also of convergence around some of the key questions which have shaped the postmodern debate; forms of convergence which *do* allow us to attempt a preliminary mapping of the contributions. The various chapters thus move from perspectives on the restructuring and reconceptualization of urban spaces within political-economic as well as cultural globalization, to reconceptualizations of the notion of globalization itself as a metaphor of the changing relations between the global and the local, to a critique of the very fundamentals of the cartographic reason which has forged the modern vision of the world (and thus also the global/local dichotomy).

I have chosen to group the chapters into three sections – though not because the sections exhaustively delimit convergences between theoretical approaches, as they are surely a gross oversimplification of the distinctiveness of the individual contributions. This shorthand divide is intended, rather, as one possible reading of the debate on a postmodern geographical praxis. In fact, the divisions that I have traced also have their outliers: as the reader will note, the volume opens with Michael Dear's paper which I have chosen to inaugurate the debate for its comprehensive overview of the postmodern turn in the Anglo-American academy. The final section is followed, on the other hand, by

Gunnar Olsson's exploration of the results of the postmodern debate in geography thus far, calling to task *all* of the praxiological perspectives presented within the pages of this book.

* * *

The first chapter in the volume – Michael Dear's *The Postmodern Turn* – offers, as I noted above, a broad overview of the past years of the postmodern debate in geography and its political-strategic implications. Following an examination of some of the most prominent attacks against postmodernism articulated by critics on both the political right and left, Dear goes on to query the diverse impacts of postmodern thought on human geography and its sister discipline of urban planning. He notes that while human geography has enjoyed a remarkable efflorescence as a consequence of its critical engagement with postmodernism, the impact of postmodernism on the discipline of urban planning has been almost negligible; according to Dear, a silence attributable to the clientelism which continues to dictate planning dialogue and which precludes significant engagement with critical theory. The disciplinary consequences of such a divergence are not only of enormous importance in and of themselves but, perhaps even more vitally, serve to illustrate the extent to which "postmodernism matters" in our intellectual and political lives. As the author argues, any postmodern disciplinary praxis is, at base, "about standards, choices, and the exercise of power – in real life and the academic world. It places the construction of meaning at the core of any social theory . . . The key issue is authority. And postmodernism has served notice on all those who seek to assert or preserve their hegemonic position in academic and everyday worlds." It is just this challenge which grants postmodern critique its weight: "in individual lives, in disciplinary fortunes, in academic privilege and power, in culture wars, postmodernism matters" – as it does in the practice of geography.

Cities . . .

Dear's reflections lead us into the first section of the book which centers, broadly, upon the restructuring of contemporary urban spaces – albeit the contributions contained within it (Soja, Mitchell, and Katz) present widely divergent positions regarding the efficacy and role of postmodern interpretations in reading these transformations. This is certainly the most explicitly political section of the volume and perhaps also the most genuinely Anglo-American in tone, not only because of the empirical reference-points which guide the authors' reflections (the American city in general, and Los Angeles and New York in particular), but also because the contributions offer reflections of the more generalized debate of the past years within Anglo-American geography on the emergence and meanings of the postmodern city.

The section opens with Edward Soja's 'Exploring the Postmetropolis,' which summarizes the author's approach to critical urban studies and extends his past work on postmodern urbanism. Soja argues forcefully that there have been profound changes taking place in what we familiarly describe as the modern metropolis: changes which necessitate a significant modification of the ways in which we understand, experience, and study cities. To distinguish "what has changed the most from what remains most constant and continuous," Soja adopts the term "postmetropolis" and outlines the past thirty years of urban restructuring as defining what he dubs a significant "postmetropolitan transition": "a selective deconstruction and still ongoing reconstitution of the modern metropolis as it has traditionally been understood and interpreted." A transition which, according to the author, implies the emergence of forms of "postmodern urbanism"; of a novelty of approaches to the critical study of cities and regions, "informed by a variety of post-prefixed modes of interpretation: poststructuralist, post-Fordist, post-colonial, post-Marxist, postmodern feminist." The chapter thus offers a broad overview of the theoretical and political implications of changes to the "modern" urban form over the past decades, arguing, however, that any such changes cannot be understood in unequivocal fashion, for "there are no purely postmodern cities or completely postmodern urban regions, and perhaps there never will be," as "continuities with the past, at the very least with distinctively modernist forms of urbanism and urbanization, persist everywhere, shaping the spatial, social, and historical dimensions of contemporary urban life." Yet the ambiguity or extent of what Soja terms the "postmodernization" of cities does in no way diminish its importance (both in theoretical as well as practical terms) in making sense of the contemporary city and ongoing urban restructuring processes. According to Soja, it is only a critical postmodern geographical praxis, capable of negotiating the complex co-presence of modernist and postmodernist interpretative perspectives and approaches, that can provide more open and flexible understandings of the postmetropolitan transition, thus facilitating a progressive place-based spatial politics.

The chapter that follows is Don Mitchell's 'Postmodern Geographical Praxis? Postmodern Impulse and the War Against Homeless People in the "Post-Justice" City,' which explicitly challenges the emancipatory potential of postmodern urbanism lauded by Soja and Dear. Mitchell contends, in fact, that geographers' (and others') utopian fantasies about a supposed postmodernization of the city serve to sustain not a progressive urban politics but, rather, regulatory regimes that are increasingly brutal in intent and practice: "not some revolutionary nomadism but the control over mobility in the name of 'order' and 'civility.'" The "actual spaces" of the postmodern city "as it is being built in bricks, mortar and law" are, according to Mitchell, increasingly circumscribed within a narrowly defined order. For the author, postmodern urbanism represents, above all, a privatized urbanism, with "whole realms of social reproduction auctioned off wholesale to private police forces, prison operators, parks conservancies, business improvement districts and the like," and ideals of a public/common good pushed aside in favor of "a punitive form of urbanism meant only to protect the interests of the privileged few." Mitchell's chapter focuses, in particular, on the evolution of this shift in American cities over the past decade, noting especially the emergence of a spate of new laws designed to control, regulate, and usually criminalize the spaces and everyday behavior of homeless people. What the author argues, through an analysis of the proposals for public space zoning, is that they are, at least potentially, genocidal for homeless people, with the conceptual tools of postmodern urbanism providing "exactly the language needed to effect, in law and in design, the elimination not of homelessness but of homeless people." Mitchell's arguments thus question the liberatory potential of a postmodern geographical praxis: according to the author, one which, "as it unfolds in the academy, has become little more than a handmaiden to the brutal postmodern geographical praxis of the streets – despite all intents to the contrary."

Mitchell's critique of postmodern urbanism is extended by Cindy Katz in 'Hiding the Target: Social Reproduction in the Privatised Urban Environment.' Within this chapter, the author develops the notion of a "hidden city of social reproduction," elaborating the ways in which the uneven relations and material practices of social reproduction are respectively hidden and targeted by the neoliberal urban agenda. In particular, Katz focuses upon the Grand Central Partnership in New York City and its pursuit of the neoliberal urban vision through the rhetoric of preservation, noting how such rhetoric (and the ensuing practice) has acted to remove from view particular actors and their social activities, "in the interests of ensuring 'orderly,' 'clean,' and 'safe' public space." The chapter thus highlights the ways in which the forces behind the Partnership have come together "to deter further corporate flight from Manhattan, and to revitalize midtown in ways that would draw new forms of investment and bring middle- and upper-class visitors – local and not – back to New York." A strategy whose success, according to Katz, should be seen as embedded in the much broader "political economic context of reinvestment in New York by major financial institutions; the restructuring of global capitalism in ways that produced and relied upon a few high-technology financial markets or 'global cities,' among them New York; and the emergence of neoliberalism, associated with massive public disinvestment in social spending and the privatization of all manner of public and municipal services." All marks, according to the author, of the "postmodernization" of the city; all marks which highlight "concerns with visibility and invisibility associated with power, wealth, and poverty." Katz thus calls for geographers to recuperate the presence, work, and struggles of those who are effaced from the preserved landscapes of "architecturally 'significant' buildings and militaristic monumentality": to "scratch at the surface of the contemporary city to uncover working landscapes not represented in traditional geographies, architectures, histories of place, to counter the politics enabled by the erasure of particular constituencies from the built environment." It is precisely such efforts, according to the author, which can "disturb the comfortable illusions of a beneficent globalized capitalism"; which can lay down the foundations of a truly progressive geographical praxis which, as Katz concludes, can "lead someplace far more important than a commercial mezzanine."

Scales . . .

Within the second section of the volume, I have grouped contributions which seek to problematize the notion of scales of analysis in the examination of the processes of globalization. All three chapters – albeit informed by widely diverse theoretical frameworks and examining an accordant diversity of empirical 'objects' – come together around a common concern: that of highlighting the highly normative nature of such taken-for-granted categories as the global and the local, conceived as unitary and separate scales of geographical analysis within which to conceptualize today's socio-economic, cultural, and territorial processes. Each of the authors (Dematteis, Flusty, and Smith) within this section, in fact, offers suggestions for alternative readings which could escape from this false dualism – as well as from the scalar strictures of 'modern' geography more broadly.

The first chapter within this section is Giuseppe Dematteis's 'Shifting Cities,' which presents an examination of the functional and physical "shifts" in the urban form and the emergence of increasingly complex and flexible urban orders. What Dematteis argues, however, is that what geographers have taken to dubbing the "postmodern" city can, actually, be better interpreted as simply a "shifting" modern city. These shifts in the modern city can be identified and interpreted at a number of levels, according to the author. Most broadly, Dematteis identifies a shift of the urban scene emergent from the post-Fordist transition, economic globalization, environmental change, and the incipient new world political order: a shift towards more complex and flexible spatialities but which, Dematteis argues, is still geared towards maintaining and reinforcing a modern capitalist order. A second shift is apparent in the "form of the city," both physical as well as functional: a shift from the premodern model of the *urbs-civitas* spatial coincidence and modern conurbation images, city-regions, and urban fields towards new conceptual images of a spatial networking continuum clustered around the scattered fragments of an exploded centrality (exemplified by the Italian notion of the '*città diffusa*,' the French '*périurbanisation*,' or American 'edge cities'). Finally, such shifts in the spatial form are also accompanied by fundamental shifts in geographical images of the city – and thus changes in its very meaning. However, as Dematteis notes, there exists a growing gap between spatial representations of the city and interpretations of urban processes, for the paradigms of neoclassical geography founded on relationships of physical proximity are no longer capable of representing the changes wrought by today's processes of globalization. What is needed, according to Dematteis, is a new, complex praxis of geography: a geographical praxis able to represent hyperconnected, hypermodern urbanization, and to design truly postmodern urban conditions. In the conclusions of the chapter, the author thus proceeds to outline some steps towards the elaboration of such a praxis as a complex, co-evolutive though often conflictual process of interaction between global forces, local actors, and local milieux.

Steven Flusty's 'Adventures of a Barong: A Worm's-Eye View of Global Formation' similarly calls upon geographers to counter the erasure of subjects from urban discourse and practice, calling for an "em-placed" understanding of another conceptual instrument which guides present-day geographical analyses: that is, globalization. As Flusty notes, globalization is an increasingly omnipresent notion, invoked as the underlying dynamic in a host of real-world occurrences ranging from worldwide financial crises to neighborhood demographic change. But what is globalization, exactly? Numerous competing theories of globalization have emerged over the past decade, yet despite their variety, as Flusty notes, all appear to privilege structural levels of analysis, for even when efforts to "bring globalization down to earth" are undertaken, such efforts commonly entail "a scale of resolution no finer than aggregate data on migratory demographics or capital circulation." As Flusty stresses, such privileging of "higher order" manifestations positions everyday minutiae as symptomatic of globalization, rather than as productive of it, and represents the spaces in which the everyday occurs as abstracted substrates or as entirely annihilated by technological mediation. To counterbalance this tendency, what are sorely needed are "conceptions of globalization looking to specific *in situ* everyday activities as not merely informed by the global, but fundamentally formative of the global." Flusty's chapter thus proposes an alternative approach to globalization by tracking the course of a single artefact, a men's dress shirt from southeast Asia, and its passage between individual members of numerous social groups as they negotiate their daily lives within and between the neighborhoods of Los Angeles; an approach which, as the author notes, "proceeds from the assumption that globalizing processes are human endeavors and, thus, may be fruitfully analyzed from the starting point of the personal and the particular." What Flusty suggests, then, is a conceptual and theoretical model of globalization which does not derive from conceptions of "the world as a whole" but that, rather, builds outward from the spatially situated interactions of concrete persons. Globalization may be inscribed within flows of capital and waves of migration, within satellite broadcasts and transoceanic air-routes. But, as Flusty stresses, "the global is no less in the heads and commonplace interactions of those whose lives underpin these larger-order phenomena." All views of the global, he concludes, are "views from the inside. We are the capillaries of globalization, variously participating in complex webs of emerging relationships that are simultaneously spatially extensive and psychically intensive. We actively produce the 'local holes' through which the 'global flows' we hear so much about percolate or, more precisely, are percolated." To grasp the "global," Flusty argues, geographical praxis must turn to its "most intimate bases" – those of localized everyday existence.

Neil Smith's 'Rescaling Politics: Geography, Globalism and the New Urbanism' concludes this section of the volume. Moving from the meditations of a sixteenth-century Venetian cartographer striving for an ideal inscription of the world, to the "scale bending" pretenses of New York City Mayor Rudolph Giuliani, Smith argues that it is only through an historicized examination of the "archeology of scale" that we can reveal the politics of social difference, and begin to "reconnect the postmodern with the modern"; in a sense, then, speaking to the underlying questions of scale framing all the chapters within this section. Recent years, as Smith's chapter notes, have witnessed a very real shift in the scales of global power: a shift which has accompanied the concurrent

wholesale geographical restructuring of the world and which profoundly "jars our sense of scalar propriety – every social function at its
proper spatial scale – highlighting the largely hidden work that geographical scale does in ordering and maintaining our assumptions of
sociopolitical normality." Smith points out that the whole range of
postwar assumptions legislating the appropriate behaviors and functions of specific actors (state, corporate, urban, private) at specific scales
is fast dissolving, leading to the formation of new scalar assumptions;
new scalar assumptions which act to inscribe social differences in new,
diverse ways into the geographical landscapes of the city; "produced
geographies which legislate, with greater or lesser success, specific
assumptions about social and political difference and right." As Smith
stresses, scale is a "social technology, but insofar as it is also a political
repository of assumed social difference rights, it is simultaneously a
deeply disguised ideological expression – the 'representational site' – of
ruling social ideas." Yet although geographical scale may be a means of
political containment, it can also provide a means of social and political empowerment, with sociospatial struggles and political strategizing
often centered precisely upon issues of scale. In the final portion of the
chapter, Smith turns to an examination of the question of scale within
the rhetoric of the "new urbanism" which, he argues, is principally characterized by a dramatic restructuring of the relationships between economic production, social reproduction, and political governance within
cities: a "restructuring of geographical scale which represents a class-,
race-, and gender- specific remapping of the known world." Geographers' challenge in deciphering such remappings lies precisely in their
ability to theorize the social production of geographical scale; "refilling
the 'content' of social space with its ideologically 'emptied matter.'"
This, for Smith, is the essence of any progressive geographical praxis,
for "insofar as scale is the socially devised metric of spatial differentiation, it is difficult to conceive of a geographical project that is more
acutely political."

Mappings . . .

The third and final section of the volume expands the reflection on the
essentially cartographic global readings of the world highlighted to one
extent or another by both Smith and Dematteis and looks to alternative (postmodern?) mappings of the world – and of geography and geographers' roles within it. Denis Cosgrove and Luciana de Lima Martins'
call for new "performative mappings" is followed in this section by the
reflections of three Italian geographers (Minca, Guarrasi, and Farinelli),

all centring on a critique of cartographic logic as the mainstay of modern geography. As all of the above authors argue, a postmodern geographical praxis can only emerge by revealing the "original act of forgetting" which underpins positivist geography; a forgetting which has allowed modern geography to mistake an ontological problem for an epistemological one, thus masking the contradictions and paradoxes inherent within it and transforming fundamental questions of knowing and representing the world into but 'technical' problems.

The first chapter in this section is, as I noted, Denis Cosgrove and Luciana de Lima Martins' 'Millennial Geographics,' focusing upon millennial celebrations in London and Rome and examining the ways in which millennialism and performances of the millennium can be seen as emblematic of the current opening of geographical discourse and praxis towards the creative arts. Cosgrove and Martins conceptualize celebrations of the millennium as acts of "creative globalization" that aim to universalize the Western/Christian space-time underpinning dominant expressions of the geographical imagination. Yet in order to be enacted, they note, the meanings of millennialism (death and renewal, origins and ends, memory and desire) must not only be associated with a specific calendrical moment, but must also be mobilized at particular locations across the globe; a process which seeks to actively make and remake historic *genius loci* in determinate locations through a whole series of spatial interventions and performative events which might be regarded as characteristically postmodern in content and style. The authors' examination of such millennial "mappings" thus begins by tracing the contradictions between the universalism and implicit time-place fixity of the idea of the millennium on the one hand, and the polyvocality of postmodern conceptions of time-space on the other. Cosgrove and Martins' reflections, however, also speak to the implications of non-essentialist notions of space and, in particular, of the uses of a "performative mapping" for geographical praxis in the twenty-first century: a "performative mapping" mirroring the contemporary convergence between artistic and scholarly activity in geography, "simultaneously understanding and activating spatialities." The authors argue that it is only by admitting the inescapable ambivalence of all processes of geographical representation that geographers can "move beyond sterile science-versus-humanities debates within geography, reflecting much broader epistemological shifts apparent at the millennium," and towards "an active engagement in the processes of knowledge-making in which we are immersed, a movement that not only historicizes our understanding of scientific practices, but also emphasizes the creative and imaginative dimensions of scientific activity."

My own chapter ('Postmodern Temptations') attempts an exploration of recent critiques of the paradoxes and contradictions inherent in 'modern' ways of conceptualizing space and describing the world; in other words, the crisis of geographical representations. After tracing the outlines of the disciplinary debate which, over the course of the past decade, has attempted to lay down the bases for the construction and legitimation of an "other" ("postmodern"?) geographical praxis able to transcend the closed spaces framed by modern cartographic logic, I elaborate some suggestions for an alternative praxis of geography capable of "traveling beyond the map." Grounding my reflections within the theoretical considerations of Timothy Mitchell and, in particular, his critique of the "metaphysics of representation" developed in the pages of *Colonizing Egypt*, I query the possibilities of a geographical praxis which frees itself from its time-honored search for the true and eternal meaning of places; an understanding/reading of places that is partial and positioned, but always open since infinitely recontextualizeable. A geographical praxis, then, as a tool through which we can both narrate the world as well as construct it – in a continual and shifting process of negotiation and articulation with a multiplicity of other praxes.

Vincenzo Guarrasi's considerations on the 'Paradoxes of Modern and Postmodern Geography: Heterotopia of Landscape and Cartographic Logic' carry on the examination of the role of the literary and the performative highlighted by Cosgrove and de Lima Martins in chapter 8 by turning to what the author terms "cartographical irony" as a means of confronting the spatial metalanguage of modernity. Guarrasi begins his analysis with an examination of the concept of landscape as a metalinguistic act; one of the multiplicity of metalinguistic acts that allow us to codify and recodify the world around us. According to Guarrasi, landscape, in fact, constitutes the archetypal heterotopia of modernity, inscribing and circumscribing the world or, more accurately yet, modernity itself; thus, "to exit from the confines of modernity, one must somehow suspend, neutralize, or invert the notion of landscape"; one must "suspend belief in its very existence." Of the rhetorical figures which we adopt in order to grant meaning to the world (such as metaphor, synecdoche, and metonymy, following Duncan 1990), it is only irony, according to Guarrasi, which has not been transposed into cartographic terms. For Guarrasi, it is the contextualized nature of irony that allows it to "reveal" subjectivity and to unmask the play of rhetoric. Therefore, if rhetoric (and the rhetoric of landscape in particular) leads us to what Meyer (1997) terms the "willing suspension of disbelief," Guarrasi argues that irony can perhaps allow for the opposite effect: a sort of "willing suspension of belief" in the cartographic logic of modernity – or in landscape itself. Guarrasi proposes that one such possibility is presented precisely by geographical information systems, the *ratio extrema* of cartographic logic; GIS which, as digital representations of landscapes, renders explicit the distance between the modern subject and the world, revealing the rhetorical devices of landscape and thus "suspending belief" – allowing, in other words, for a "postmodern" reading of space.

The last chapter of this section, Franco Farinelli's 'Mapping the Global, or the Metaquantum Economics of Myth,' turns to quantum theory as a possible source of inspiration for a postmodern geographical praxis. Farinelli begins his account by dissecting the "cartographic dictatorship" that has framed modern geographical thought since Ptolemy. He argues that modern (or "classical") geography has always been predicated upon an implicit silencing of the relationship between subject and object, between the observer and the object of observation; a relationship which has always been mediated by the cartographic apparatus. Noting recent attempts by geographers to write the complexity of today's globalizing world, Farinelli points out that modern cartographic understandings are, by their very nature, unable to model a world which "no longer functions as a map." In order to begin to trace truly global geographies, Farinelli argues that geographers should turn to the "messy power of complexity" afforded by recent reflection in quantum theory. It is quantum theory, he argues, that forced classical physics to face the crisis of the validity of the cartographic model and of its "tabular logic." A similar reflection is long overdue in geography, particularly as applied to that mainstay of the cartographic method that is projection: an operation whose rules "geographers may know well, but whose deeper nature they entirely ignore." By applying the basic principles of quantum logic, Farinelli argues that geographers could be led to "remember" the relationship between the viewing subject and cartographic representation and thus transcend the insurmountable distance that, in modern geography, has separated subject from object. Quantum understandings provide but a preliminary opening to the formulation of new, postmodern geographies, however. What Farinelli suggests is that it is through myth and myth-making that geographers can apply the principles of quantum theory to that which he terms "the physics of the world." It is the pre-cartographic nature of myth and of mythical ways of knowing that renders them so precious in today's global world. The ontology of myth, as Farinelli argues, "precedes all dissociation, its ways of knowing precede any category," and it is only in myth that "we can conceive of a world not composed of a seriality of separate individuals, each with his/her distinct place and thus exclusive one of another . . .; the world of myth, rather, admits the existence of subjects who co-participate, subjects who are mutually interpenetrating." It is myth, then, that can allow for a postmodern geography able to fashion pre/post-cartographic, "global" understand-

ings of the world, geographies able to describe the world not as a map
but as a sphere.

*   *   *

If we can suggest that Michael Dear's opening chapter provides the
stable context for the Anglo-American 'wing' of the postmodern debate,
then the role of Gunnar Olsson's closing commentary 'Washed in a
Washing Machine' is certainly that of destabilizing *all* of the contributions to this volume, by pointing to the underlying self-referential
problem of all 'postmodern' geographies. Pausing upon René Magritte's
Ceci n'est pas une pipe, Olsson proceeds to outline the binary divide
born of cartographical reason which, he argues, continues to mark geographical perspectives: modes of understanding which "simultaneously
tells us where we are and where we should go." The challenge of a truly
postmodern geographical praxis, according to Olsson, is to move
beyond the "grayish space which fills the void between the picture of
the pipe and the story of the words"; a challenge "not to write about
something [but] to produce a text which at the same time is that something." Starting from the art of Marcel Duchamp and his revolutionary
experiments with the taken-for-granted (and progressing to the work of
a number of contemporary artists), Olsson elaborates a critique of the
supposed "postmodernism" of self-professed postmodern geography,
challenging geographers, rather, to practice "a postmodern geography
as intense as Mark Rothko's paintings, so closely tied to the prohibited"; a postmodern geographical praxis which leaves "no limit sacred,
no member untouched" and which has the courage to challenge "the
interface of the I and the Thou."

*   *   *

A few brief final considerations, then. Although the structuring of the
volume into sections certainly guides readings of the chapters along
three distinctive interpretative trajectories, it is important to keep in
mind that *all* of the contributions speak (to one extent or another) to a
common problem: that of the conflation between theoretical reflection
and empirical research, between description and prescription. And, I
would argue, it could not be otherwise, for the postmodern turn in
geography has not only paved the way for reflection on the constitution

of new urban spatialities, but has also elicited new explorations of the politics inherent within all of the categories that we, as geographers, adopt to describe and narrate the world.

Our readings of the world are, inevitably, normative practices that codify, that create the organization of the world itself; and it is within this codification, within this creative force that geographical metaphors reveal all of their power. If there exists, indeed, one characteristic uniting the diversity of approaches from which to interrogate a possible post-modern geographical praxis, it is certainly their *common awareness of the normative (and not merely descriptive) power of modern geo-graphical metaphors; their awareness of the invisible mechanisms of modernity that, precisely through a strategic use of spaces and spatial representations, succeed in masking their inherently violent logic.*

My not-so-secret hope is that this volume can furnish a series of alter-native readings of the (post)modern but, above all, that it can hint at the possibilities of new geographical practices able (through a decon-struction and a condemnation of the taken-for-granted) to reclaim a political and strategic dimension; new postmodern geographical prac-tices free of the self-referential indulgences and linguistic games which expose them to ready attacks from the 'practitioners' of geography – from those who claim to 'do' geography in their decision-making and through a naturalization of their seemingly endless repertoire of the taken-for-granted.

I hope that within these chapters the reader may find a variety of *other* ways of 'doing geography'; other ways of doing geography that demand a full responsibility, that require the researcher to fully render explicit her or his distinct positionality, *ad nauseum*, if necessary. All ways of doing geography, however, that despite their partiality never renounce their political and strategic task; geographies that are aware that they do not speak of truths but, rather, that they articulate posi-tions and judgments according to determinate value systems (always rendered explicit and open to confrontation with those of others) as, I believe, the authors of this volume have had the courage to do.

In the name of all the homeless forcibly removed from the new urban spaces, in the name of all geographical imaginations that refuse to be subdued within linear logics, in the name of all of those who do not admit the death of all projects other than the local, in the name of all of the lived geographies of the everyday hidden if not entirely erased by hegemonic discourses, in the name of all of those who refuse the social determinism of the market – in the name of all of these things, I believe it is our duty as geographers to reflect on the possibilities as well as the risks inherent in a postmodern praxis of geography.

And although such a new praxis is sure to be articulated in a multiplicity of languages emergent from the spaces hidden or erased by the modern project, in a seemingly cacophonous rebellion to the closed and inherently violent spaces of cartographic logic, we should not fear its diversity. In all of its tongues it must continue to cry out that the King is naked, it must continue to deconstruct all of the trappings of the taken-for-granted. For it is precisely in the struggle to expose that which is taken-for-granted that a postmodern praxis can find the will to reimagine its spaces, can begin to map out new social coordinates, can give birth to new strategies that give space to forgotten voices.

1

The Postmodern Turn

Michael Dear

The term *postmodern* is used promiscuously to refer to a multitude of ideas, but I believe that there are three principal references in post-modern thought: (1) a series of distinctive cultural and stylistic practices that are in and of themselves intrinsically interesting; (2) the totality of such practices, viewed as a cultural ensemble characteristic of the con-temporary epoch of capitalism (usually distinguished by the term *post-modernity*); and (3) a philosophical and methodological discourse antagonistic to the precepts of Enlightenment thought, most especially the hegemony of any single intellectual persuasion. In this chapter, I shall engage all three ideas – postmodernism as style, epoch, and philoso-phy/methodology.

One of the most intriguing aspects of the postmodern term is the intensity of the attacks against postmodernism. I begin by examining some prominent examples of this vitriolic assault, from political per-spectives on the right and the left. Then I shall explore postmodernism's impact on two closely related disciplines (human geography and urban planning) to reveal enormous differences in the consequences of the dis-ciplines' engagement with postmodern thought. To anticipate my direc-tion, I shall conclude that the academic discipline of human geography has enjoyed a remarkable efflorescence as a consequence of this engage-ment, but the professional orientation of urban planning has (for clearly discernible reasons) almost totally ignored its challenges.

Against Postmodernism

'And in the Right corner...'

Post-modernism entices us with the siren call of liberation and creativity,
but it may be an invitation to intellectual and moral suicide.

Gertrude Himmelfarb

Gertrude Himmelfarb, an eminent and conservative historian, attacked
postmodernism in a long article in the *Times Literary Supplement* in
1992. It is a classic of its kind. First she claims that postmodernists tell
'good historians' nothing that they do not already know:

> Modernist history is not positivist in the sense of aspiring to a fixed, total,
> or absolute truth about the past. Like post-modernist history, it is
> relativistic, but with a difference, for its relativism is firmly rooted in
> reality. It is sceptical of absolute truth but not of partial, contingent, incre-
> mental truths. More important, it does not deny the reality of the past
> itself.... [The] modernist historian reads and writes history... with a
> scrupulous regard for the historicity, the integrity, the actuality of the past.
> He makes a strenuous effort to enter into the minds and experiences of
> people in the past, to try to understand them as they understood
> themselves, to rely upon contemporary evidence as much as possible, to
> intrude his own views and assumptions as little as possible, to reconstruct
> to the best of his ability the past as it 'actually was'... (Himmelfarb
> 1992: 12)

Himmelfarb recognizes the impossibility of this ideal, conceding that
historians have known all along that their work is vulnerable on three
counts: "the fallibility and deficiency of the historical record on which
it is based; the fallibility and selectivity inherent in the writing of history;
and the fallibility and subjectivity of the historian" (Himmelfarb 1992:
12). The strength of good history relies on a tried and tested historical
method that favors objectivity, but is contemptuous of a postmodernism
that engages with the contingency of historical explanation:

> Critical history put a premium on archival research and primary sources,
> the authenticity of documents and reliability of witnesses, the need for
> substantiating and countervailing evidence; and, at a more mundane level,
> on the accuracy of quotations and citations, prescribed forms of docu-
> mentation in footnotes and bibliography, and all the rest of the 'method-
> ology' that went into the 'canon of evidence'. The purpose of this
> methodology was twofold: to bring to the surface the infrastructure, as
> it were, of the historical work, thus making it accessible to the reader and
> exposing it to criticism; and to encourage the historian to a maximum

exertion of objectivity in spite of all the temptations to the contrary. Post-modernists scoff at this as the antiquated remnants of nineteenth-century positivism. But is has been the norm of the profession until recently. (Himmelfarb 1992: 12)

The fallibility and relativity of history is not, in Himmelfarb's eyes, a discovery of the postmodernists. The difference between the old and the new lies in the presumptions of postmodernism: that "because there is no absolute, total truth, there can be no partial, contingent truths . . . [and] that because it is impossible to attain such truths, it is not only futile but positively baneful to aspire to them" (Himmelfarb 1992: 12).

In postmodernism's repudiation of Enlightenment principles, Him-melfarb identifies a political agenda that is subversive of the social order:

> Post-modernism is . . . far more radical than either Marxism or [other] new 'isms' . . . all of which are implicitly committed to the Enlightenment principles of reason, truth, justice, morality, reality. Post-modernism repu-diates both the values and the rhetoric of the Enlightenment. In rejecting the 'discipline' of knowledge and rationality, post-modernism also rejects the 'discipline' of society and authority. And in denying any reality apart from language, it aims to subvert the structure of society together with the structure of language. (Himmelfarb 1992: 14)

Even in its celebration of difference, postmodernism threatens the intel-lectual order: "The modernist accuses the post-modernist of bringing mankind to the abyss of nihilism. The post-modernist proudly, happily accepts that charge" (Himmelfarb 1992: 15).

Paradoxically, while she concedes the political posture implicit in any historical account, Himmelfarb decries our descent into a politics of recognition, which, she claims, has removed all meaning from history:

> Multiculturalism has the obvious effect of politicising history. But its more pernicious effect is to demean and dehumanise the people who are the subjects of history. To pluralise and particularise history to the point where people share no history in common . . . is to deny the common (generic) humanity of all people, whatever their sex, race, class, religion, and the like. It is also to trivialise history by so fragmenting it that it lacks all coherence and focus, all sense of continuity, indeed, *all meaning*. (Himmelfarb 1992: 14; emphasis added)

Needless to say, Himmelfarb judges that young people are particularly at risk of contamination from postmodernism, but she is hopeful at the first signs of disaffection with postmodernism, because the appeal of novelty will surely recede:

The 'herd of independent minds' . . . will find some other brave, new cause to rally around. Out of boredom, careerism (the search for new ways to make a mark in the profession), and sheer bloody-mindedness (the desire to *épater* [skewer] one's elders), the young will rebel, and the vanguard of today will find itself an aging rearguard . . . (Himmelfarb 1992: 15)

While unclear about whether or not the era after postmodernity will restore the old, or usher in some even more despicable fad, Himmelfarb concedes that the traditional verities of the history profession will unlikely be recovered intact.

What Himmelfarb appears to fear most is the loss of the historical method that lies at the heart of her vision of the discipline. It is noteworthy that she equates methodological weakness with moral failing:

One can foresee a desire to return to a more objective and integrated, less divisive and self-interested history. What will be more difficult to restore is the methodology that is at the heart of that history. A generation of historians . . . lack any training in that methodology. They may even lack the discipline, *moral as well as professional*, required for it. (Himmelfarb 1992: 15; emphasis added)

The traditional history now being discarded is hard and exciting, according to Himmelfarb, involving truth, objectivity, coherence, accuracy, and good footnotes:

the old history, traditional history *is* hard. Hard – but exciting precisely because it is hard. And that excitement may prove a challenge and inspiration for a new generation of historians. It is more exciting to write true history (or as true as we can make it) than fictional history, else historians would choose to be novelists rather than historians; more exciting to try to rise above our interests and prejudices than to indulge them; more exciting to try to enter the imagination of those remote from us in time and place than to impose our imagination upon them; more exciting to write a coherent narrative while respecting the complexity of historical events than to fragmentize history into disconnected units; more exciting to try to get the facts (without benefit of quotation marks) as right as we can than to deny the very idea of facts; even more exciting to get the footnotes right, if only to show others the visible proof of our labours. (Himmelfarb 1992: 15)

It remains unclear exactly how Himmelfarb proposes to achieve her true history, rise above her own prejudices, excise her own imagination, prevent herself from imposing narrative coherence where none may exist, decide what 'facts' matter, and ensure that her footnotes sufficiently attest to her serious commitment to a hard history.

John M. Ellis, in *Against Deconstruction* (1989), arrives at broadly similar conclusions. In his case, it is deconstructionists who offer nothing new to literary critics:

> The only sense in which deconstruction can be said to represent change in the critical context lies in its giving new shape and a renewed force and virulence to pre-existing ideas and attitudes: it has given an appearance of theoretical sophistication to what had previously been the more or less incoherent attitudes and prejudices of majority practice. (Ellis 1989: 153)

I am a little leery about reading this too literally. Ellis appears to be saying that the incoherent practices of the majority of his peers have been normalized by deconstructionism, a situation he laments. But what is clear to Ellis is that some people want to reject "standards of intelligent criticism" with consequences "well-nigh everyone" (presumably a different majority than the one just identified?) deplores:

> The point of good criticism cannot lie in its discovering *the* meaning of a text; to use that criterion would be to return to the unitary truths of science . . . The prevailing critical consensus, then, has long insisted on pluralism, on the value of different viewpoints . . . If no one cares to talk about standards of intelligent criticism, then the content of published criticism will vary enormously – and it does; well-nigh everyone, including the most ardent advocates of theoretically unrestrained pluralism, is unhappy with at least some aspects of this situation. (Ellis 1989: 154, 155, 156)

Ellis recognizes a yet deeper problem in "that deconstruction's major themes are themselves *inherently antitheoretical* in nature" (Ellis 1989: 158).The process by which he reaches this conclusion is revealing (158–9). First, he asks us to understand that theory is, axiomatically, disruptive:

> Theory exerts its pressure on the status quo by continual examination of the basis and rationale for the accepted activities of a field of study. Inevitably, the results will in principle be quite unlike the usual deconstructive attitudes: they should not leave us with issues and activities more undefined than ever but instead introduce a clarification and differentiation of fundamentally different kinds of activities. The consequence of this kind of activity will generally be a pressure to rearrange priorities; *by its very nature, then, theory is indeed disturbing.* (Ellis 1989: 158; emphasis added)

But the kind of theory that Ellis sanctions has little room for unfettered freedom of thought, even as it appears to encompasses many of the practices which would meet the approval of those who practice (say) 'hard' history:

> But theoretical argument must . . . proceed with great care. It must be above all a careful, patient, analytical process: its strengths must lie in precision of formulation, in well-drawn distinctions, in carefully delineated concepts. In theoretical discourse, argument is met by argument; one careful attempt to analyse and elucidate the basis of a critical concept or position is met by an equally exacting and penetrating scrutiny of its own inner logic. That makes theoretical argument very much a communal process: there is *no room in it for individual license*, for claims of exemption from logical scrutiny, for appeals to an undefined unique logical status, for appeals to allow obscurity to stand unanalysed, or for *freedom to do as one wishes*. (Ellis 1989: 158–9; emphases added)

The 'most enduring fault' of deconstructionist literary criticism is exposed in Ellis's own *credo*:

> A shared inquiry means a commitment to argument and dialogue, while a criticism that insists on the value of each individual critic's perspective, in effect, refuses to make that commitment. Before deconstruction, theory of criticism worked against the laissez-faire tendencies of criticism; but now deconstruction, an intensified expression of those tendencies, has attempted to seize the mantle of theory in order to pursue this antitheoretical program. The result is an apparent novelty that, looked at more closely, consists in resistance to change and, more particularly, to that change that is most urgently needed: the development of some check on and control of the indigestible, chaotic flow of critical writing through reflection on what is and what is not in principle worthwhile – that is, through genuine, rather than illusory, theoretical reflection. (Ellis 1989: 159)

Ellis is unclear about who gets to decide what is 'worthwhile,' or what are the standards of 'genuine' theoretical reflection. I suspect it will be those same people who decided which standards were important before postmodernism came along.

'And in the Left corner . . .'

> Postmodernity is merely a theoretical construct of interest primarily as a symptom of the current mood of the Western intelligentsia.
>
> *Callinicos 1990: 9*

One of the more plain-spoken attacks lobbed from the Left comes from Alex Callinicos in his unambiguously titled volume: *Against Postmod-*

ernism: A Marxist Critique (1990). His opening sentences reveal that he, too, has contracted the virus that seems endemic among anti-postmodernists: irritation.

> Yet another book on postmodernism? What earthly justification could there be for contributing to the destruction of the world's dwindling forests in order to engage in debates, which should surely have exhausted themselves long ago? My embarrassment in the face of this challenge is made all the more acute by the fact that at the origins of the present book lies that unworthy emotion, irritation. (Callinicos 1990: 1)

I suppose this is meant to be engaging; certainly it's intended to trivialize. In any event, it does not hold out much hope for a tolerant appraisal on the part of the Irritated Intellectual. And sure enough, Callinicos promptly and unambiguously denies any merit in postmodernism as philosophy, style, or epoch. Letting the boundaries between poststructuralism and postmodernism slither through his fingertips, he loudly declares:

> I deny the main theses of poststructuralism, which seem to be in substance false. I doubt very much that postmodern art represents a qualitative break from the Modernism of the early twentieth century. Moreover, much of what is written in support of the idea that we live in a post-modern epoch seems to me to be of small caliber intellectually, usually superficial, often ignorant, sometimes incoherent. (Callinicos 1990: 4–5)

Callinicos could have stopped there, and himself have saved a few trees from clear-cutting. After that tirade, there can be little doubt about where he stands, can there? Well, but then he adopts what turns out to be a fairly common strategy among opinionated critics: an immediate qualification that, actually, there are some good ideas in postmodernism after all. Callinicos squirms:

> I should, however, make a qualification to the judgement just passed. I do not believe that the work of the philosophers now known as post-structuralist can be dismissed in this way: wrong on fundamentals Deleuze, Derrida, and Foucault may be, but they develop their ideas with considerable skill and sophistication, and offer partial insights of great value. (Callinicos 1990: 5)

I am sure that Deleuze, Derrida, and Foucault will be relieved. But Callinicos never really engages with which of these poststructuralist theses are 'false' and which of 'great value.' Callinicos' main gripe is the way in which postmodernism deflects attention from a "revolutionary socialist tradition" (Callinicos 1990: 7). His own intellectual project:

rather uneasily occupies a space defined by the convergence of philosophy, social theory and historical writing. Fortunately, there is an intellectual tradition which is characterized precisely by the synthesis it effects of these genres, namely the classical historical materialism of Marx himself, Engels, Lenin, Trotsky, Luxemburg and Gramsci. (Callinicos 1990: 5–6)

The fact that there may be other intellectual traditions, which address this synthesis, is of no consequence to Callinicos:

Only, I contend, classical historical materialism, reinforced by an account of language and thought that is naturalistic as well as communicative, can provide a secure basis for the defense of the 'radicalized Enlightenment' to which Habermas is committed. (Callinicos 1990: 7)

This is an odd world toward which Callinicos entices us. There is no need for an alternative epistemology; assume Marxism, and everything will fall into place.

Callinicos on the Left, like Himmelfarb from the Right, warns of the dire, politically disastrous consequences of pursuing the pied pipers of postmodernism. As before, he offers us a stark (crudely oversimplified) vision of politics, showing few signs of empathy with the changing world about him :

Unless we work toward the kind of revolutionary change which would allow the realization of this potential in a transformed world, there is little left for us to do, except, like Lyotard and Baudrillard, to fiddle while Rome burns. (Callinicos 1990: 174)

Terry Eagleton's *The Illusions of Postmodernism* (1996) offers another Marxist-inspired attack, although it is curiously adumbrated in its scope. Eagleton concedes that while he finds the distinction between postmodernism and postmodernity useful, "it is not one which I have particularly respected in this book" (Eagleton 1996: viii). In addition to lethargy on this key topic, he also betrays advanced symptoms of the 'straw man' syndrome, confessing that he is less concerned with "the higher philosophical flights" of postmodernism, but more with:

what a particular kind of student today is likely to believe; and though I consider quite a lot of what they believe to be false, I have tried to say it in a way that might persuade them they never believed it in the first place. (Eagleton 1996: viii)

Quite what this indigestible mouthful is meant to convey remains a mystery to me. Eagleton seems willing to deal only with bastardized caricatures of popularized arguments that may be invented by impressionable young minds – hardly an invitation to a sophisticated, sustained engagement!

In the event, as one might expect from him, Eagleton's critique is wily and witty, betraying the author's weakness for wanton wordplay. He wends a now-familiar way through a blistering criticism to another of those contrite confessions that postmodernism does after all have some merit, some of the time. At the end of the book, a kind of balance sheet appears (Eagleton 1996: 143):

> [The] rich body of [postmodern] work on racism and identity, on the paranoia of identity-thinking, on the perils of totality and the fear of otherness: all this, along with its deepened insights into the cunning of power, would no doubt be of considerable value. But its cultural relativism and moral conventionalism, its scepticism, pragmatism, and localism, its distaste for ideas of solidarity and disciplined organization, its lack of any adequate theory of political agency: all these would tell heavily against it.

In a review of Eagleton's book, Ian Pindar laments the fact that philosophers are now being blamed for the failure of politics. Pindar claims that Eagleton and others are simply shooting the messenger: "there can be few who seriously believe that this [failure] is because our politicians have read too much Baudrillard" (Pindar 1997: 25).

Not all Left critics use the apocalyptic tones of Callinicos and Eagleton. In *What's Wrong with Postmodernism* (1990), Christopher Norris invokes the usual mantra that postmodernism offers nothing new. He even traces a proto-postmodernism in Marx's writings, endorses the opinion that a postmodern pluralism offers little beyond a politics of the status quo (Norris 1990: 33). He concludes that the main lesson to be learned from (for instance) Baudrillard's texts is:

> that any politics which goes along with the current postmodern drift will end up by effectively endorsing and promoting the work of ideological mystification. (Norris 1990: 191)

And in his rather suspiciously-titled *The Truth about Postmodernism* (1993), Norris elaborates upon the political paralysis that accompanies postmodern indeterminacy:

> Postmodernism can do nothing to challenge these forms of injustice and oppression since it offers no arguments, no critical resources or validating grounds for perceiving them as inherently unjust and oppressive. (Norris 1993: 287)

Kevin Robins provided a beautiful thumb-nail sketch of Norris's position that I cannot better. Writing with reference to the aftermath of the Gulf War, Robins says this of Norris (Robins 1996: 323):

> Against the 'intellectual fad' of postmodernity, Norris wants to vindicate, and to re-ground, what he calls 'enlightenment truth-seeking discourse.' It is [Norris claims] 'the issues of real-world truth and falsehood that provide the only basis for reasoned opposition on the part of conscientious objectors.' Postmodernism is presented as a kind of propaganda and misinformation campaign within the intellectual world. It is as if the perversion of communication and the confusion of reason could be laid at the door of this thing called postmodernism. And Baudrillard is the Great Satan.

Robins makes clear that Baudrillard's sin is "to distance himself from the cause of Reason" (Robins 1996: 323). Once again, critics are drawing a bead on the messenger.

Cartographies of complaint

The case against postmodernism may be mapped into six basic complaints, as follows.

There is nothing new about postmodernism
Critics revel in the claim that all the positions staked out in postmodernism have been present for indefinite periods of time in their respective disciplines. But his can be true only at an extraordinarily high level of generality. And in any event, should we therefore rest content with the familiar? Is there *nothing* new in the present concatenation of events and ideas? If not, why are critics so, as they put it, *irritated*?

No new era is presaged by postmodernity
The epochal dimension of postmodern thought primarily exercises those on the Left who, presumably, have most to gain by identifying historical progress toward some revolutionary millennium. Their assessment is invariable: that capitalism is alive and well, and postmodernity merely a blip on the evolutionary horizon. But this is entirely the wrong question. Epochal change need not imply the end of capitalism (think of the shift from an agricultural to an industrial capitalism). The advent of a postcolonial era is already a prominent political and epistemological feature of the late twentieth century; so is the shift to an 'information age.' A revived political economy could help us visualize shifts that the classical model hides from view.

Old traditions and standards are best
Those who defend old traditions and existing canons, or seek to reconstruct the project of Modernity, point witheringly to postmodernism's
notorious faddishness, its proliferation of meanings, and evidence of
poor scholarship. Then they steer us cantankerously back to former
headings and steadfast moorings, to canons that keep us in line and
them in control. Their favorite word is *discipline* (of the old ways). Yet
because something is fashionable, i.e. of its time, does not mean it is
wrong. If we multiply meanings and confusions, it's the fault of our
minds. (We are like aging gymnasts pulling on the stretched spandex of
modernism, watching as its meaning becomes more threadbare with
each application.) And, let's face it, poor scholarship is everywhere; it
is certainly not the prerogative of postmodernists.

Postmodernism's relativism has produced a cacophony of competing
voices that are difficult to distinguish
This is the complaint of a lazy and hostile critic. Relativism is OK, so
say the keepers of the keys, as long as it stays in its place. But as soon
as existing authority is threatened, relativism becomes an issue (just as
it is now), and counterattacks are launched. The person of Comfortable
Authority has no incentive to seek an accommodation with other voices;
the easier option is to dismiss them, even as the volume of contemporary voices testifies to the depth of silence imposed by earlier hege-
monies. These voices are also proof-positive of the plurality of
interpretative possibilities, none of which can claim to be authoritative
or self-evident.

Postmodernism is antitheoretical/atheoretical
The pejorative terms anti- or a-theoretical are usually invoked when the
critic's own dearly-held beliefs are being questioned. For instance, Ellis
insisted on well-drawn distinctions, exacting and penetrating scrutiny
of internal logic, and a communal process; Himmelfarb required coherence, consistency, factuality; and so on. Such critics seem constitutionally incapable of conceding that alternative criteria are possible for
judging scholarship. Paul Feyerabend among others has demonstrated
that antitheoretical and atheoretical stances have an important and
honorable status in scientific inquiry.

Postmodernism is bad politics
This complaint has two parts: first, plain revulsion because postmodernism undermines all ideologues (rabid Republicans and dismal
Democrats of all stripes); and second, impatience because a postmodern political agenda is slow in evolving. But while it may be difficult to

envisage a postmodern politics, this in no way denies the relevance of the postmodern condition to contemporary politics. Postmodernists cannot be blamed for the failures of the body politic, whose wounds have been largely self-inflicted.

In my terminology, these cartographies of complaint can be summed up as follows:

- postmodernism as style is simply an ephemeral fad of no lasting value;
- postmodernism as method cannot undermine centuries of Enlightenment wisdom; and
- postmodernism as epoch has no purchase since capitalism still exists.

All three positions are easily refuted. Postmodern style was once a new fashion, but it has survived for several decades; indeed it has itself been canonized, and it cannot be ignored by those who profess to learn from the contemporary. Second, I have no doubt that the conventions of 'science' will lead us to important, life-saving, and life-threatening discoveries; but I am equally convinced that the way we know things has been irrevocably altered by a postmodern consciousness. (Can anyone countenance cloning with equanimity, or have faith in the benign reassurances of scientists?) And thirdly, capitalism may be alive and well, or it may be dying. No one can tell. But the accumulated weight of change in the latter part of the twentieth century suggests to me that epochal change is in the making. I wish a long life to those who deny this. May we all live long enough to see clearly what is already plain on the face of the earth.

Transformations in Geography and Planning

A very revealing difference is observable between the disciplines of geography and urban planning; the former has clearly absorbed many postmodern traditions and been altered by them; the latter has all but ignored postmodernism. What has caused this difference? And with what consequences? These are questions of professional and academic politics, as much as they reflect intellectual differences. My explanation must inevitably be preliminary, because we remain perforce caught up in the postmodern turn; it will focus less on detailed contributions than on broad intellectual trends; and (at the risk of causing offense) I shall be naming many scholars who probably would not label themselves as postmodernists because their works are (to me) clearly implicated in the advent of postmodernism.

Postmodern geographies

The tidal wave of postmodernity hit human geography with predictable consequences. As in many other disciplines, it engendered intense excitement in a handful of scholars inspired by its provocations. But more generally, it met with active hostility from those who perceived their intellectual authority being threatened; incomprehension on the part of those who (for whatever reason) failed to negotiate its arcane jargon; and the indifference of the majority, who ignored what they presumably perceived as simply the latest fad. Despite the combined armies of antipathy and inertia, postmodernism has flourished.

The year 1984 is significant because it was then that Fredric Jameson published what many regard as *the* pivotal English-language article focusing geographers' attention on spatiality and postmodernity. Ten years later, Jameson's essay retained its vitality, and human geography had undergone a revolution of sorts. Two of the earliest geographical articles that took up Jameson's challenge were by Dear and Soja (Dear 1986; Soja 1986). The former dealt with urban planning; the latter was an exuberant deconstruction of Los Angeles by an avowed postmodernist. Both articles appeared in a special issue of the journal *Society and Space* devoted to Los Angeles.[1] Between 1986 and 1994, over 50 major articles and an equivalent number of critical commentaries have appeared in prominent geography journals including especially *Society and Space*, but also the *Annals of the Association of American Geographers*, the *Canadian Geographer*, and the *Transactions of the Institute of British Geographers*.[2]

Traces
With the benefit of hindsight, traces of a postmodern consciousness can, of course, be uncovered in geographical writings prior to 1986 (Gregory 1989).[3] The principal historical reasons for the absorption of postmodern thought into geography are properly to be found in the resurgence of Marxist social theory in the late 1960s and 1970s. (See for example King 1976; Gould 1979.)[4] It was out of a broadly-based poststructuralist response to the perceived obsolescences of Marxism that impetus was imparted to the postmodern turn. In geography, this trend was instrumental in the renaissance of interest in social theory more generally, thus reconnecting the discipline with a broad spectrum of socioeconomic and political debates. Of particular consequence were the substantive emphases on the urban question (Castells 1977), and the role of space in economic development and socio-spatial relations (Harvey 1982; Massey 1984). The relatively high levels of scholarly pro-

ductivity and output in these areas rendered them particularly suscep-
tible to innovation and rapid evolution.

It was not long before the neo-Marxist revival fell under scrutiny, and
something like a 'golden age' of theoretical/philosophical efflorescence
occurred in human geographical thought.[5] For instance, in 1978 Derek
Gregory published his influential *Science, Ideology and Human Geography*, drawing attention in particular to the work of critical theorists
such as Jürgen Habermas. In that same year, Gordon Clark and I began
our reappraisal of the theory of the state, with a poststructural emphasis on the languages of legitimacy (Dear and Clark 1978). A humanist
geography also developed to counter the Marxian emphasis (Ley and
Samuels 1978), and during this period very deliberate attempts were
launched to investigate the ontological and epistemological bases of geographical knowledge. This was manifest in, for instance, Ed Soja's determined efforts to reposition space in the realm of social theory (Soja
1980); in Gunnar Olsson's (much-maligned) confrontation with language (Olsson 1980); and in Andrew Sayer's (1974) clear-sighted
inquiry on method in the social sciences (Sayer 1984).

The burgeoning connections between geography and social theory
were given concrete expression in 1983, with the appearance of the
journal *Society and Space* as part of the *Environment and Planning*
series. The first issue included Nigel Thrift's wide-ranging reformulation
of the problematic of time and space (which reflected his earlier work
with Allan Pred on time geography) (Thrift 1983), and Linda McDowell's fundamental paper on the gender division of urban space
(McDowell 1983). Subsequent issues have maintained a steady flow of
increasingly self-conscious attempts to link social theory and human
geography. The ubiquity of this problematic may be gauged from the
title of an influential 1985 collection of essays: *Social Relations and
Spatial Structures* (Gregory and Urry 1985). The 1986 papers by Soja
and Dear may thus have crystallized a pervasive turbulence in geography's theoretical discourse and provided a platform for the next stages
in the conversation. Yet these essays were not so much theoretical departures, but more a culmination of a decade's reengagement with the
central issues of social theory.

Consciousness
Jameson's identification of architecture as the privileged aesthetic of a
postmodern culture made it easy for geographers to adapt his insights
to their spatial agendas. Early studies by Ted Relph and David Ley drew
attention to the semantics of the built environment in the landscapes of
postmodernity (Relph 1987; Ley 1987). These and other studies were
instrumental in provoking an uninterrupted sequence of research on

postmodern culture, emphasizing place and place-making, spectacle and carnival, and consumption. It was also inevitable that cultural geographers would be drawn to postmodernism. An early appraisal of the 'new' cultural geography is provided by Peter Jackson (1989) (see also: Agnew and Duncan 1989; Cooke 1988a; Cosgrove and Daniels 1988; Larkham 1988; Mills 1988; Sack 1988; Shields 1989). An independent line of geographical inquiry in the late 1980s centered on the processes of contemporary economic restructuring, particularly the move toward flexible specialization (what some call flexible accumulation). Economic geographers were attempting to analyze the emergent dynamics of post-Fordist, flexible industrial systems and their concomitant spatial organization. Although few if any of these inquiries were explicitly postmodern in nature, they inevitably intersected with the problematic of periodization, i.e. whether or not a radical break had occurred to signal the arrival of a postmodern society (Cooke 1988b; Gertler 1988; Schoenberger 1988; Scott 1988; Storper and Walker 1989). A third source of fertile intellectual discord concerned the emergent status of social theory in human geography. The validity of a social theoretical approach was rarely at issue; more usually, the debate took the form of sometimes vitriolic exchanges among competing orthodoxies, the details of which need not detain us here (see Saunders and Williams 1986).[6] A temporary truce established two broad positions: one coalition favored maintaining the hegemony of their preferred theory (whatever that happened to be); a second advocated a theoretical pluralism that may properly be viewed as a precursor of postmodern sensibilities.

The point that these trends establish is that a postmodern consciousness emerged in human geography not from some orchestrated plot, but instead from a diversity of independent perspectives – including cultural studies, emergent economic geographies, and stand-offs in social theory.[7] Each trend had a life of its own before it intersected with postmodernism, but each (I believe) was irrevocably altered as a consequence of this engagement. By 1988, the climate was such that I was able to argue for what I styled the 'postmodern challenge' in human geography (Dear 1988). My plea was premised on the significance of space in postmodern thought and the potential of geography's contribution to a rapidly evolving field of social inquiry.

The wave
The year 1989 saw the publication of two books with postmodern geography as a central theme. Soja's *Postmodern Geographies* was a celebration of postmodernism and its challenges; Harvey's *The Condition of Postmodernity* was a hostile critique of postmodernism that attempted to subsume it within the explanatory rubric of Marxism. A

year later, Phil Cooke's *Back to the Future: Modernity, Postmodernity and Locality* appeared – a perspective on the 'localities' project in Great Britain which was sympathetic to the claims of postmodernism. Whatever their respective merits, these books and their authors concentrated a discipline's attention on the postmodern question.[8] But in truth, the wave had already gathered an unstoppable momentum. The roster of publications in 1989 and subsequent years reveals a burgeoning postmodern consciousness in the three topical areas I previously identified:

1 *cultural landscapes and place-making*, with an increasing emphasis on the *urban* (Anderson and Gale 1992; Beauregard 1989; Dear 1989; Duncan 1990; Glennie and Thrift 1992; Hopkins 1990; Robins 1991; Shields 1989; Short 1989; Zukin 1991; Barnes and Duncan 1992; Sorkin 1992; Dear and Wolch 1989);
2 *economic landscapes of post-Fordism and flexible specialization*, with particular interest in *global–local connections and the spatial division of labor* (Barnes and Curry 1992; Dunfor 1990; Gertler 1988; Leborgne and Lipietz 1988; Sayer and Walker 1992; Schoenberger 1988; Scott 1988; Slater 1992a, 1992b; Storper and Walker 1989; Webber 1991);
3 continuing *philosophical and theoretical disputes*, especially those relating to *space* and the problems of *language* (Curry 1991; Doel 1992; Folch-Serra 1989; Hannah and Strohmayer 1991; Harris 1991; Jones, Natter and Schatzki 1993; Milroy 1989; Peet and Thrift 1988; Philo 1992; Pile 1990; Schatzki 1991; Scott and Simpson-Housley 1989; Smith 1989).

There was also an explosion of interest in the application of postmodernism to other areas, representing a deepening appreciation of the extent of postmodernism's reach and relevance.[9] The many themes that became manifest during the period 1989–93 may be grouped under four broad rubrics:

1 *problems of representation in geographical/ethnographic writing* (Barnes and Duncan 1992; Crang 1992; Jackson 1991; Marcus 1992; Matless 1992a; Katz 1992; Keith 1992; Reichert 1992), in *cartography* (Harley 1989; Pickles 1992; Wood 1992), and in *art and film* (Bonnett 1992; Daniels 1992; Aiken and Zonn 1994);
2 the historical and contemporary *politics of postmodernity* (Dalby 1991; Driver 1992; Graham 1992; Hepple 1992; O'Tuathail 1992; Pile and Rose 1992), *feminist geography*'s discontentment with postmodernism (Bondi and Domosh 1992; Christopherson 1989; Domosh 1991; Pratt 1992), *orientalism and postcolonialism* (Driver

1992; Gregory 1991), and the *law* and critical legal studies (see Blomley and Clark 1990);

3   an emphasis on the *construction of the individual and the boundaries of self*, including *human psychology and sexuality* (respectively, Bishop 1992; Hoggett 1992; Geltmaker 1992; Moos 1989; Knopp 1992; Valentine 1993);

4   a reassertion of *nature and the environmental question* (Bordessa 1993; Emel 1991; Fitzsimmons 1989; Matless 1991, 1992a, 1992b), which has taken many forms, including a fresh look at the relationships between *place and health* (Gesler 1993; Kearns 1993).

By 1991, postmodernism had received an extended treatment in a textbook on geographical thought (Cloke, Philo, and Sadler 1991), and became part of the standard fare in others (see for example Johnston 1991; Livingstone 1992). Matters were further helped by the publication of some important works in English translation (Lefebvre 1991). The availability of Lefebvre's *La Production de l'Espace* was especially important.

In pedagogic terms, postmodernism's emphases on difference and diversity have penetrated the academy, but its impact is difficult to document.[10] I have no idea how much postmodernism is actually being taught in geography departments, but attempts to reconcile cultural and social geography may be regarded as a step in this direction (Philo 1991). A number of institutional responses in the late 1980s and early 1990s reflected a growing awareness of the need to dissolve disciplinary barriers in teaching and research. For instance, a Center for Critical Analysis of Contemporary Culture was set up in 1986 at Rutgers University, since when Neil Smith and other geographers have played an important role in a broadly-based social science and humanities research program. In 1989, at the University of Kentucky, a Committee on Social Theory was founded in order to encourage campus-wide collaboration, again with a strong organizational base in geography (including John-Paul Jones and John Pickles). And an interdisciplinary master's degree in Society and Space admitted its first students in 1992 at the University of Bristol in England, based in the Department of Geography and the School of Advanced Urban Studies.

Contentions

The introduction of postmodernism into human geography was not without dissent. The most common complaints echoed those already current in the intellectual marketplace: that postmodernism's extreme relativism rendered it politically incoherent, and hence useless as a guide for social action; that it was itself just one more metanarrative; and that the project of modernity remained relevant even though there was little agreement about exactly which pieces were worth salvaging. I have also already noted feminism's divergent path.

At a superficial though certainly not trivial level, many geographer critics simply lost patience with the promiscuous way in which the term had been bandied about; if it could be applied to everything, then it probably meant nothing. Others were upset that they and their work were invoked to support a movement for which they had no sympathy.[11] In one such case, Allan Pred angrily distanced himself with these words:

> I have never chosen to label myself as "postmodern". . . . I regard "postmodern" as an inaccurate, uncritical, deceptive, and thereby politically dangerous "epochal" labeling of the contemporary world . . . [which is] best depicted as modernity magnified, as modernity accentuated and sped up, as *hyper*modern, not *post*modern. (Pred 1992)

This is an unequivocal rejection of the postmodern, even though Pred's work is clearly implicated in the rise of postmodern geography. The most sustained rejection of the postmodern turn in geography was undoubtedly that of Harvey (1989). Given his unassailable reputation within and beyond the discipline, it was to be expected that the book would be widely read and the repudiation it contained would deal a mortal blow. But, while broadly acknowledged, the book did little to stall the production of postmodern geographical scholarship. The fact that the book met with some stinging rebuttals may have muted its influence within the discipline (for example Dear 1991; Massey 1991). In addition, Harvey's orthodoxy might have posed problems for fellow Marxists who had begun the long and arduous task of rewriting their theory to account for the altered conditions of postmodernity.[12]

A different literature was less concerned with outright rejection of postmodernism and more with a critical engagement with its problematic. Most commonly, this work explored the genealogy of postmodern thought, its broad links with the modern era, and the persistence of modernist themes in the present discourse (Curry 1991; Strohmayer and Hannah 1992). Postmodern thought invigorated an effort to define the parameters of modernity itself (for example, Ward and Zunz 1992; see also Giddens 1990). Julie Graham perceptively examined the consequences of postmodernism for a progressive politics (Graham 1992). Finally, some geographers joined the push to go beyond the terms of the current debates (Pile and Rose 1992; Thrift 1989, 1993; see also Borgmann 1992).

Postmodern planning

[W]e can be sure that theories of modernity and postmodernity do not help us much here; indeed, they represent a rather huge red herring in our trawl through the theoretical waters. Even if they were clearer and more coherent and more consistent, which they quintessentially are not, they deal only with a very small slice of historical reality . . . And they are deliberately, rather infuriatingly, aspatial: they are entirely uninterested in the question of what happens where, and why.

Hall 1988: 14

By the mid-1990s, the impact of postmodern thought on the theory and practice of city planning was almost negligible. In some respects this is unsurprising since the precepts of postmodernism seem exactly anti-thetical to the rationalist foundations of urban interventionism. More-over, it is unlikely that any clients in public or private sectors would want to see undermined the rationality/expertise bulwarks of planning, nor have their own power position exposed. On both counts, one can perhaps forgive planners their quiescence when confronted with the counter/narratives of postmodernism. But I still want to know why there is this apathy.

In 1986 I published "Postmodernism and Planning," which was, I believe, the first article to link the two subjects. In the years since then, I have counted about a dozen articles in major journals that directly address at least some aspect of postmodernism. Some of them have been written by the same authors, so the actual number of contributors to the debate is even less. The earliest of these contributions were essen-tially enthusiastic acceptances and extensions of the original arguments (see Simonsen 1990; Punter 1988; Milroy 1989, 1990, 1991). Most notable were several pieces by Beth Moore Milroy, who intelligently pursued the implications of a changing social context for emergent prac-tices of planning. By the early 1990s, the discourse was still limited. Dennis Crow made an intriguing connection between postmodernism and the earlier modernisms of Le Corbusier's *The City of Tomorrow and its Planning*. His was an important reminder of the contingent, con-tinuing relevance of the project of modernity.

About the same time, Bob Beauregard began a series of insightful essays looking beyond obsolete modernist codifications of planning. He left no doubt that postmodernism had undermined the intellectual authority of the modernist planning project, which was consequently

suspended between a modernism whose validity is decaying and recon-figuring, and a postmodernism whose arguments are convincing yet dis-comfiting. As planning theorists we have failed to formulate a response and failed to work with practitioners to move the planning project from its ambivalent position. (Beauregard 1991; see also 1989)

Through an evocative critique of Peter Marris's novel *The Dreams of General Jerusalem* (a story of urban planning in an African postcolonial regime), Beauregard elaborates on this perceived failure:

> first, modernist planners have lost touch with the prevailing political-economic forces that are restructuring cities and regions in a global context, and second, have failed to keep pace with concomitant intellectual currents and cultural forms. (Beauregard 1991: 192)

From this condemnation, Beauregard goes on to take some important steps toward a postmodern ethos: that knowledge is not necessarily a reliable guide to action, and that increased understanding is likely to reveal more differences rather than set directions. But yet, he asserts, "Action can be unequivocal, knowledge can be helpful, and people can struggle successfully to improve their lives" (Beauregard 1991: 193). And, in a later study, he argues in favor of an ideal city which is "democratic, egalitarian, multi-racial and non-sexist" (Beauregard 1994: 6). Complaining about the demise of utopian thinking in planning, and the cooptation of the profession by private capital, Beauregard urges planners to engage in public debates about the 'good society' and its attendant physical form. The normative core of such discourse should be "social and spatial justice and empowerment" for "fair procedures and . . . more just distributions" (Beauregard 1994: 12). But, he warns, in discussing alternative urban futures, "we must avoid lapsing into utopian fantasy or stifling multiple voices" (Beauregard 1994: 12). This is where Beauregard falters, in my judgment. He seems to be issuing a contradictory mandate here, specifying the need for dialogue about social justice in a good society, but warning that such dialogue should avoid too much utopianism or too much polyvocality. But how much is enough? If this is an attempt to envision a way out of the postmodern impasse, then Beauregard has bought his way out at a high price – *viz.*, the silencing of visionaries and minorities. This can hardly be what he intended, but this is the paradoxical drift of his advice.

Beauregard is alert to the issues and manages a sustained involvement with them; others have been less successful. For instance, in 1992 Charles Hoch used insights from postmodern and critical theory to argue that practicing planners have hidden behind a cloak of professional competence, and they lack the capacity to identify or acknowledge their use of power. He shows how the "postmodern critique leaves no place for planning to hide from its attachment to power" (Hoch

1992: 207). And he calls for the establishment of open "moral communities" that cut across established lines of bureaucracy and community affiliation.[13] So far so good. Two years later, however, Hoch published a book entitled *What Planners Do: Power, Politics, and Persuasion* (1994). It is undoubtedly the most authoritative text to date on this topic. Yet in it, Hoch makes not one mention of postmodernism, nor the ramifications of his conclusions of two years previously. Instead, he calls on the planner to use "craft" and "character" to build a "reform community" (Hoch 1994: ch. 12).

No such ambivalence is present in Harper and Stein, who in the mid-1990s presented what was then the most sustained engagement with the challenge of postmodern thought in planning (Harper and Stein 1995). The title of their article makes their position clear: "Out of the Postmodern Abyss: Preserving the Rationale for Liberal Planning." And, lest there be any doubt about where they stand, they write in their introduction:

> Our intent is practical. Our concern is not that planning theory might be flawed by some technical philosophical error, but rather that the uncritical adoption of postmodernist assumptions would bring us to the brink of an abyss of indeterminacy, impairing our ability to maintain social continuity through change, to treat each other in a fair and just and fully human way, and to justify public planning. (Harper and Stein 1995: 233)

These are serious charges (but where *does* this paranoia about uncertainty come from?!) though Harper and Stein never truly demonstrate what it is about postmodernism that would lead to a revolutionary, unjust, inhuman, and planning-less society. Instead, they advocate a form of *neopragmatism* as a "coherent and reasonable basis for justifying planning" (Harper and Stein 1995: 233). Placing *dialogue* as the heart of any model of postmodern planning, Harper and Stein concede the positive implications of postmodernism: the "rejection of metanarrative," the "distrust of rigid methodology," the "celebration of plurality," and the recognition that "all voices have the right to be heard" (Harper and Stein 1995: 240). But they warn:

> full-blown [*sic*] postmodernism cannot provide an adequate basis for planning. . . . Our concern is that, taken to its extreme, full-blown postmodernism would inevitably reduce planning to the impotent state . . . the full-blown postmodernist alternative seemingly leaves no room for planning.[14]

They're probably right. It explains why planners tiptoe gingerly around postmodernism as if it were some form of deadly disease (full-blown, at that), which shouldn't be discussed, in polite company. The more extreme forms of postmodernism would certainly imply the end of all disciplines and professions.

To save planning from the apocalypse, Harper and Stein urge upon us a neopragmatism that is (in their words) nonfoundational, anti-essentialist, neither absolutist nor relativistic, fallibilistic, nonretractive, and nonscientific. They claim it will produce a more powerful "understanding and critique of planning practice" (Harper and Stein 1995: 241), an assertion based on the following slogans (among others):

1   "Persuasion through rational argument is the only alternative to power" (Harper and Stein 1995: 241).
2   "A neopragmatic perspective allows for . . . objective critique" (Harper and Stein 1995: 242).
3   "we can have a planning that is truly rational" (Harper and Stein 1995: 242).
4   "neopragmatism frees us to do what we know we *should* do" (Harper and Stein 1995: 242; emphasis in original).

By adopting a neopragmatic planning, Harper and Stein conclude:

> Planners can get what they want from postmodernism (a broader notion of rationality, recognition of multiple voices and discourses, inclusivity, encouragement of many voices, empowerment) and retain significant aspects of modernism (emancipation, accountability, hope for the future). (Harper and Stein 1995: 242)

It is easy to raise an eyebrow at Harper and Stein's outrageously starry-eyed elisions over the palpable contradictions of their reconstructive vision. But they do not lack courage or conviction, and they have dealt seriously with postmodernism.

It was left to Bob Beauregard to sustain the discipline's interest in things postmodern during the mid-nineties. In his highly original 1993 study of the rhetoric of decline in the history of US cities (including the discourse of planners), Beauregard's intent is to "convey the discourse as it was heard and read by people who lived during those times" (Beauregard 1993: x). In a direct assault on the tension between interpretative strategies and 'objective' analyses, he mixes methodologies lasciviously and confronts head on the question of representation:

> My intent is to subvert the authority of the discourse and compel the reader to confront its indeterminacy. (Beauregard 1993: 49)

Beauregard maintains this stance until the very last chapter of the book when he at last allows himself to enter directly into the analysis in order to probe "the ideological core of the discourse in order to reveal its foundational dynamics" (Beauregard 1993: 280). He concludes:

> There will never be one true story of urban decline, but that should not discourage us from attempting to pin down its meanings or from pursuing coherent and compelling interpretations. (Beauregard 1993: 281)

This is very compelling. More than anyone else in the profession, Beauregard was/is, I think, actually practicing a version of postmodern planning.

I thought Beauregard was alone until an important new collection of essays was published in 1996 under the title *Explorations in Planning Theory* (edited by Mandelbaum, Mazza, and Burchell). In one of the essays, our old nemeses Harper and Stein continue their 'scorched earth' policy, this time attacking the incommensurability premise in comparative theoretical analysis. They begin their essay with what seems an indefensible statement: "The influence of postmodernism on planning theory seems to be increasing" (Harper and Stein 1996). As before, they are concede the merit of some of postmodernism's charges; but once again they rally round reason and the liberal paradigm to save the day. The most interesting thing about this book is that several contributors are actually talking about postmodernism, if only indirectly. (It may be significant that many of these voices are those of non-Americans, and many are female.) For instance, here is Jean Hillier taking a "postmodern and feminist" sensibility to the discourse of planning; Judith Allen conjoining with Foucault to examine knowledge-based politics in London; and Bent Flyvbjerg on the dark side of modernity (Hillier 1996; Allen 1996; Flyvbjerg 1996). Now, the revolution has not yet arrived; there are about a score more essays in the publication that do not broach the P word.

Most recently, Leonie Sandercock published her *Toward Cosmopolis: Planning for Multicultural Cities* (Sandercock 1998). I was then, and remain now, energized by this book's appearance, because Sandercock's vision of a 'postmodern utopia' is a brilliant demonstration of what happens when planning truly engages postmodernism, postcolonialism, and feminist thought. While her book and mine proceed from similar concerns, her emphasis is much more on urban planning practice *per se*. In it, she is highly critical of the conventional, modernist pillars of planning wisdom, instead favoring an inclusive, people-centered style of planning that emphasizes practical wisdom, multiculturalism, and community empowerment (Sandercock 1998: 27–30). She unabashedly aims for a "postmodern Utopia," which can never be realized, but will "always be in the making" (Sandercock 1998: 163). It deals with "social justice, a politics of difference, new concepts of citizenship and community, and a civic culture formed out of multiple publics" (Sandercock 1998: 199). An essential ingredient of Sandercock's postmodern planning paradigm is:

> A reinstatement of inquiry about and recognition of the importance of memory, desire, and the spirit (or the sacred) as vital dimensions of healthy human settlements and a sensitivity to cultural differences in the expressions of each. (Sandercock 1998: 214)

Sandercock has the courage to dream about a postmodern utopia (Sandercock 1998: 218–19). The very existence of her book is sufficient proof, for me, of the value of the encounter with postmodernism.

The War of the Words

Whether we approve or are even aware of it, the postmodern wave has already broken over geography. Some chose to ride the wave; others ducked under, hoping that this too would pass. I believe that we have witnessed a revolution of sorts in human geographical thinking since 1984. In general terms, there has been:

1 a reassertion of the role and significance of space in social theory and social process;
2 an unprecedented rise in scholarship devoted to the relationship between space and society;
3 a reintegration of human geography with mainstream social science and philosophy;
4 a totally new appreciation of diversity and difference, and a consequent diversification of theoretical and empirical work;
5 a self-conscious questioning of the relationship between geographical knowledge and social action;
6 an enormous efflorescence of research topics and publications.

Some or all of these events may have occurred without the advent of postmodernism; but I doubt it, at least not with the same intensity and consequences.

In contrast, the impact of postmodernism in urban planning theory and practice has been minuscule. The most obvious reason for the screaming silence on postmodernism is *clientism*. Consumers of planning in public and private sectors do not want uncertainty and ambiguity; nor will they pay for therapy for professionals with theoretical angst or identity crises. Instead, they reward competence and expertise, and the manly ability to see a deal through. They don't want anyone exposing (still less questioning) the powerful who dwell behind the planning dialogue. Clientism in architecture turned postmodernism into a marketing device; practicing planners instead chose 'new urbanism' as the warmed-over marketing concept *du jour*.

Clientism does not explain the lack of academic curiosity in postmodernism, although if we call students 'clients' of planning education, their insistence (for the moment, at least) on a practice-relevant and development-oriented curriculum may play a part. Despite John Friedman's marvellous maps of the intellectual terrain occupied by planners, the fact is that planners have only ever colonized tiny pieces of that terrain, apparently suspicious of the traditions of social theory. Thus, there have been no efforts to parallel John Forester's sustained engagement with Habermas, for instance. Indeed, the prominence afforded hermeneutics in planning theory is partly a consequence of the *absence* of alternative theories in the impoverished dialectical universe of planners. Even when alternative visions have surfaced, as in the case of feminist theory, there seems to have been enormous difficulties in keeping them alive. In fairness, these are difficult times of retrenchment and recession in the academy and profession, and survival in a Darwinistic universe may be more pressing than intellectual refinement. I also understand that planners, like everyone else, cling tenaciously to their beliefs and status. Knowledge is, after all, power and we are loathe to relinquish the basis for our claims to legitimacy. But is a critical openness too much to ask for? Since comparison, analogy, and metaphor are some of the principal means by which human knowledge is advanced, it would indeed be an unusual science that refused to look tolerantly beyond the horizons of tradition, or be discomfited if others cast a critical eye in its direction.

Looking ahead, I am both optimistic and pessimistic. In one respect, John Ellis was correct in his critique of deconstructionism: that it appealed not because it was a radical departure from entrenched attitudes, but because it fitted the *already prevailing* climate of intellectual pluralism and lent that climate a new legitimacy (Ellis 1989: ch. 7). This is not a particularly surprising insight, given what we know about the situatedness of theory. What's more interesting is how Ellis declares himself unable to live with the consequent 'chaotic flow' of critical writing, how he pleads for a return to 'standards' of intelligent criticism (ignoring the fact that exactly analogous sentiments led to the original Enlightenment exclusivities). Postmodernism is about standards, choices, and the exercise of power – in real life and the academic world. It places the construction of meaning at the core of any social theory (though this is not the only concern in such theory). The key issue is authority. And postmodernism has served notice on all those who seek to assert or preserve their hegemonic position in academic and everyday worlds. These are not trivial or esoteric concerns. In individual lives, in disciplinary fortunes, in academic privilege and power, in culture wars, postmodernism matters.

NOTES

This chapter is based on a chapter from my book, *The Postmodern Urban Condition* (Oxford: Blackwell, 2000).

1   It was no accident that much of the initial impetus to a postmodern human geography derived from southern California. This was, after all, the site of one of Jameson's most provocative postmodern encounters (with the Bonaventure Hotel). In addition, Charles Jencks, the principal chronicler of the postmodern movement in architecture, was on the faculty of the University of California at Los Angeles; the humanities program at the University of California at Irvine (in Orange County) played frequent host to Derrida, Lyotard, and other luminaries; and a deliberate attempt was underway to reconceptualize late-twentieth-century urbanism under the auspices of the LA School.

2   My survey of the literature since 1984 has been confined to English-language sources; I have also deliberately excluded from consideration the vast outpouring of postmodern literature in disciplines other than geography since that date. Both strategies were adopted to contain my review within manageable proportions. One other methodological point is pertinent: I am acutely aware that, in dealing with essays and books according to their dates of publication, I am ignoring the true chronology of conception and writing. Some may regard this as a minor problem because a work must appear in print to achieve its widest impact. On the other hand, this logic skirts the undoubted influence of precirculated drafts, conference presentations, etc. Unfortunately, I know of no straightforward way to overcome this bias.

3   Gregory (1989) provides a succinct and authoritative overview of geography's external connections with political economy, sociology, and anthropology during this period.

4   This is not to suggest that there was no relevant history before the Marxist renaissance; quite the contrary. See, for example, King (1976) and Gould (1979), a caustic retrospective on the decades between 1957 and 1977. Here, I shall focus on postmodernism's principal genealogy rather than an exhaustive disciplinary history.

5   A detailed historiography of the truly exceptional period between 1965 (the year in which Haggett's *Locational Analysis in Human Geography*

was published) and 1986 (the explicit appearance of the postmodern in geography) remains to be written.

6 See Saunders and Williams (1986) and the subsequent can(n)on fire in volume 5/4 of *Society and Space.*

7 There undoubtedly were other important trends besides the three I have identified (e.g. the localities research initiative in Great Britain). I have not attempted an exhaustive review of all the threads in the postmodern web, merely to establish their critical contributory presence prior to post-modernism's appearance.

8 My own critical assessments of Soja and Harvey are to be found in ch. 9 of *The Postmodern Urban Condition.* Other extended commentaries are to be found in Deutsche (1991), Massey (1991), and Relph (1991).

9 The rush of publications in 1992 was partly due to Marcus Doel and David Matless, who assembled two remarkable issues of *Society and Space* (vols. 10/1 and 10/2) devoted entirely to the postmodern question.

10 My argument in this paragraph closely follows that in Graff (1987).

11 This is likely to be true of some of the authors I have cited in this essay. So let me repeat my earlier caveat: not all authors referred to in my dis-cussion will see themselves or their work as implicated in the postmodern turn. However, while I have no desire to foist an unwanted label on anyone, I will insist on a connection between their works and the his-toriography of this essay.

12 See, for example, the special issue of *Antipode* 21, 1989, on "What's Left to do?," pp. 81–165, especially the essays by Clark and Walker.

13 The phrase is Mandelbaum's (1988).

14 I think the terminology of full-blown postmodernism is ill advised, since the only other context where this term is in current use is in the case of AIDS/HIV (as in: full-blown AIDS).

REFERENCES

Agnew, J. and J. Duncan, eds. (1989). *The Power of Place: Bringing Together Geographical and Sociological Imaginations.* Boston: Unwin Hyman.

Aiken, S. C. and L. E. Zonn (1994). *Place, Power, Situation and Spectacle: A Geography of Film.* Lanham: Rowman and Littlefield.

Allen, J. (1996). Our Town: Foucault and Knowledge-Based Politics in London. In *Explorations in Planning Theory,* eds. S. J. Mandelbaum, L. Mazza, and R. W. Burchell. New Brunswick: Center for Urban Policy Research, 328–44.

Anderson, K. and F. Gale, eds. (1992). *Inventing Places: Studies in Cultural Geography.* New York: Wiley, Halstead Press.

Barnes, T. J. and M. R. Curry (1992). Postmodernism in Economic Geography: Metaphor and the Construction of Alterity. *Environment and Planning D: Society and Space* 10: 57–68.

Barnes, T. J. and J. Duncan, eds. (1992). *Writing Worlds: Discourse, Text and Metaphor in the Representation of Landscape.* New York: Routledge.

Beauregard, R. A. (1989). Between Modernity and Postmodernity: The Ambiguous Position of US Planning. *Environment and Planning D: Society and Space* 7: 381–95.

Beauregard, R. A. (1991). Without a Net: Modernist Planning and the Postmodern Abyss. *Journal of Planning Education and Research* 10: 190–201.

Beauregard, R. A. (1993). *Voices of Decline: The Postwar Fate of US Cities.* Oxford: Blackwell.

Beauregard, R. A. (1994). Distracted Cities. Los Angeles: UCLA Graduate School of Architecture and Urban Planning.

Bishop, P. (1992). Rhetoric, Memory, and Power: Depth Psychology and Postmodern Geography. *Environment and Planning D: Society and Space* 10: 5–22.

Blomley, N. and G. Clark (1990). Law, Theory and Geography. *Urban Geography* 11: 433–46.

Bondi, L. and M. Domosh (1992). Other Figures in Other Places: On Feminism, Postmodernism an Geography. *Environment and Planning D: Society and Space* 10: 199–213.

Bonnett, A. (1992). Art, Ideology, and Everyday Space: Subversive Tendencies from Dada to Postmodernism. *Environment and Planning D: Society and Space* 10: 69–86.

Bordessa, R. (1993). Geography, Postmodernism, and Environmental Concern. *The Canadian Geographer* 37: 147–55.

Borgmann, A. (1992). *Crossing the Postmodern Divide.* Chicago: University of Chicago Press.

Callinicos, A. (1990). *Against Postmodernism: A Marxist Critique.* New York: St. Martin's Press.

Castells, M. (1977). *The Urban Question.* London: Arnold.

Christopherson, S. (1989). On Being outside 'The Project.' *Antipode* 21: 83–9.

Cloke, P., C. Philo, and D. Sadler (1991). *Approaching Human Geography.* New York: Guilford Press.

Cooke, P. (1988a). Modernity, Postmodernity and the City. *Theory, Culture and Society* 5: 475–92.

Cooke, P. (1988b). Flexible Integration, Scope Economies, and Strategic Alliances. *Environment and Planning D: Society and Space* 6: 281–300.

Cosgrove, D. and S. Daniels, eds. (1988). *The Iconography of Landscape.* Cambridge: Cambridge University Press.

Crang, P. (1992). The Politics of Polyphony. *Environment and Planning D: Society and Space* 19: 527–50.

Curry, M. R. (1991). Postmodernism, Language, and the Strains of Modernism. *Annals of the Association of American Geographers* 81(2):210–28.

Dalby, S. (1991). Critical Geopolitics: Discourse, Difference, and Dissent. *Environment and Planning D: Society and Space* 9: 261–83.

Daniels, S. (1992). The Implications of Industry. In *Writing Worlds*, eds. T. J. Barnes and J. Duncan. London: Routledge, 38–49.

Dear, M. (1986). Postmodernism and Planning. *Environment and Planning D: Society and Space* 4: 367–84.

Dear, M. (1988). The Postmodern Challenge: Reconstructing Human Geography. *Transactions of the Institute of British Geographers* 13: 262–74.

Dear, M. (1989). Privatization and the Rhetoric of Planning Practice. *Environment and Planning D: Society and Space* 7: 449–62.

Dear, M. (1991). Review of Harvey's *The Condition of Postmodernity*. *Annals of the Association of American Geographers* 81(3): 533–39.

Dear, M. (2000). *The Postmodern Urban Condition*. Oxford: Blackwell.

Dear, M. and G. Clark (1978). The State and Geographic Process. *Environment and Planning A* 10: 173–83.

Dear, M. and J. Wolch, eds. (1989). *The Power of Geography: How Territory Shapes Social Life*. Boston: Unwin Hyman.

Deutsche, R. (1991). Boys Town. *Environment and Planning D: Society and Space* 9: 5–30.

Doel, M. A. (1992). In Stalling Deconstruction: Striking Out the Postmodern. *Environment and Planning D: Society and Space* 10: 163–79.

Domosh, M. (1991). Toward a Feminist Historiography of Geography. *Transactions of the Institute of British Geographers* 16: 95–104.

Driver, F. (1992). Geography's Empire: Histories of Geographical Knowledge. *Environment and Planning D: Society and Space* 10: 23–40.

Duncan, J. S. (1990). *The City as Text: The Politics of Landscape Interpretation in the Kandyan Kingdom*. Cambridge: Cambridge University Press.

Dunford, M. (1990). Theories of Regulation. *Environment and Planning D: Society and Space* 8: 297–321.

Eagleton, T. (1996). *The Illusions of Postmodernism*. Oxford: Blackwell.

Ellis, J. M. (1989). *Against Deconstruction*. Princeton: Princeton University Press.

Emel, J. (1991). Ecological Crisis and Provocative Pragmatism. *Environment and Planning D: Society and Space* 9: 384–90.

Fitzsimmons, M. (1989). The Matter of Nature. *Antipode* 21: 106–20.

Flyvbjerg, B. (1996). The Dark Side of Planning: Rationality and 'Realrationalität.' In *Explorations in Planning Theory*, eds. S. J. Mandelbaum, L. Mazza, and R. W. Burchell. New Brunswick: Center for Urban Policy Research, 383–94.

Folch-Serra, M. (1989). Geography and Post-modernism: Linking Humanism and Development Studies. *The Canadian Geographer* 33(1):66–75.

Freeman, M. (1988). Developers, Architects and Building Styles: Post-war Redevelopment in Two Town Centres. *Transactions of the Institute of British Geographers* 13: 131–47.

Geltmaker, T. (1992). The Queer Nation Acts Up. *Environment and Planning D: Society and Space* 10: 609–50.

Gertler, M. (1988). The Limits to Flexibility. *Transactions of the Institute of British Geographers* 13: 419–32.

Gesler, W. (1993). Therapeutic Landscapes. *Environment and Planning D: Society and Space* 11: 171–90.

Giddens, A. (1990). *The Consequences of Modernity*. Stanford: Stanford University Press.

Glennie, P. D. and N. J. Thrift (1992). Modernity, Urbanism, and Modern

Consumption. *Environment and Planning D: Society and Space* 10: 423–33.

Gould, P. (1979). Geography 1957–1977: The Augean Period. *Annals of the Association of American Geographers* 69(1): 139–51.

Graff, G. (1987). *Professing Literature: An Institutional History*. Chicago: Chicago Press.

Graham, J. (1992). Post-Fordism as Politics: The Political Consequences of Narratives on the Left. *Environment and Planning D: Society and Space* 10: 393–420.

Gregory, D. (1989). Areal Differentiation and Post-Modern Human Geography. In *Horizons in Human Geography*, eds. D. Gregory and R. Wolford. London, Macmillan, 67–96.

Gregory, D. (1991). Interventions in the Historical Geography of Modernity. *Geografiska Annaler* 73B: 17–44.

Gregory, D. and J. Urry, eds. (1985). *Social Relations and Spatial Structure*. London: Macmillan.

Hall, P. (1988). *Cities in Civilization*. New York: Pantheon.

Hannah, M. and U. Strohmayer (1991). Ornamentalism: Geography and the Labor of Language in Structuration Theory. *Environment and Planning D: Society and Space* 9: 309–27.

Harley, B. (1989). Deconstructing the Map. *Cartographica* 26: 1–20.

Harper, T. L. and S. M. Stein (1995). Out of the Postmodern Abyss: Preserving the Rationale for Liberal Planning. *Journal of Planning Education and Research* 14: 233–44.

Harper, T. L. and S. M. Stein, eds. (1996). Postmodernist Planning Theory: The Incommensurability Premise. In *Explorations in Planning Theory*, eds. S. J. Mandelbaum, L. Mazza, and R. W. Burchell. New Brunswick: Center for Urban Policy Research, 414–29.

Harris, C. (1991). Power, Modernity, and Historical Geography. *Annals of the Association of American Geographers* 81(4):671–83.

Harvey, D. (1982). *The Limits to Capital*. Chicago: University of Chicago Press.

Harvey, D. (1989). *The Condition of Postmodernity: An Inquiry into the Origins of Cultural Change*. Oxford: Blackwell.

Hepple L. W. (1992). Metaphor, Geopolitical Discourse and the Military in South America. In *Writing Worlds*, eds. T. J. Barnes and J. Duncan. London: Routledge.

Hillier, J. (1996). Deconstructing the Discourse of Planning. In *Explorations in Planning Theory*, eds. S. J. Mandelbaum, L. Mazza, and R. W. Burchell. New Brunswick: Center for Urban Policy Research, 289–98.

Himmelfarb, G. (1992). Telling It as You Like It: Post-Modernist History and the Flight from Fact. *Times Literary Supplement*, Oct. 16: 12, 14, 15.

Hoch, C. (1992). The Paradox of Power in Planning Practice. *Journal of Planning Education and Research* 11: 207–20.

Hoch, C. (1994). *What Planners Do: Power, Politics & Persuasion*. Chicago: Planners Press.

Hoggett P. (1992). A Place for Experience. *Environment and Planning D: Society and Space* 10: 345–56.

Hopkins, J. S. P. (1990). West Edmonton Mall: Landscape of Myths and Elsewhereness. *The Canadian Geographer* 34(1): 2–17.

Jackson, P. (1989). *Maps of Meaning*. London: Unwin Hyman.

Jackson, P. (1991). The Crisis of Representation and the Politics of Position. *Environment and Planning A: Society and Space* 9: 131–4.

Jameson, F. (1984). Postmodernism, or, the Cultural Logic of Late Capitalism. *New Left Review* 146: 53–92.

Johnston, R. J. (1991). *Geography and Geographers: Anglo-American Human Geography since 1945*. London: Arnold.

Jones, J. P., W. Natter, and T. Schatzki (1993). *Postmodern Contentions: Epochs, Politics, Space*. New York: Guilford Press.

Katz, C. (1992). All the World is Staged: Intellectuals and the Projects of Ethnography. *Environment and Planning D: Society and Space* 19: 495–510.

Kearns, R. (1993). Place and Health: Toward a Reformed Medical Geography. *Professional Geographer* 45: 139–47.

Keith, M. (1992). Angry Writing. *Environment and Planning D: Society and Space* 19: 551–68.

King, L. J. (1976). Alternatives to Positive Economic Geography. *Annals of the Association of American Geographers* 66: 293–308.

Knopp, L. (1992). Sexuality and the Spatial Dynamics of Capitalism. *Environment and Planning D: Society and Space* 10: 651–70.

Larkham, P. J. (1988). Agents and Types of Change in the Conserved Townscape, *Transactions of the Institute of British Geographers* 13: 148–64.

Leborgne, D. and A. Lipietz (1988). New Technologies, New Modes of Regulation. *Environment and Planning D: Society and Space* 6: 263–80.

Lefebvre, H. (1991). *The Production of Space*. Oxford: Blackwell.

Ley, D. (1987). Styles of the Times: Liberal and Neo-Conservative Landscapes in Inner Vancouver, 1968–1986. *Journal of Historical Geography* 13: 40–56.

Ley, D. and M. Samuels (1978). *Humanistic Geography: Prospects and Problems*. Chicago: Maaroufa Press.

Livingstone, D. N. (1992). *The Geographical Tradition*. Oxford: Blackwell.

Mandelbaum, S. (1988). Open Moral Communities: Theorizing about Planning within Myths about Community. *Society* 26(1): 20–7.

Marcus, G. (1992). More Critically Reflexive than Thou: The Current Identity Politics of Representation. *Environment and Planning D: Society and Space* 19: 489–94.

Massey, D. (1984). *Spatial Divisions of Labour*. London: Methuen.

Massey, D. (1991). Flexible Sexism. *Environment and Planning D: Society and Space* 9: 31–57.

Matless, D. (1991). Nature, the Modern and the Mystic: Tales from Early Twentieth Century Geography. *Transactions of the Institute of British Geographers* 16: 272–86.

Matless, D. (1992a). An Occasion for Geography: Landscape, Representation, and Foucault's Corpus. *Environment and Planning D: Society and Space* 10: 41–56.

Matless, D. (1992b). A Modern Stream: Water, Landscape, Modernism, and Geography. *Environment and Planning D: Society and Space* 10: 569–88.

McDowell, L. (1983). Toward an Understanding of the Gender Division of Urban Space. *Environment and Planning D: Society and Space* 1: 59–72.

Mills, C. A. (1988). Life on the Upslope: The Postmodern Landscape of Gentrification. *Environment and Planning D: Society and Space* 6: 169–89.

Milroy, B. M. (1989). Constructing and Deconstructing Plausibility. *Environment and Planning D: Society and Space* 7: 313–26.

Milroy, B. M. (1990). Critical Capacity and Planning Theory. *Planning Theory Newsletter* 4: 12–18.

Milroy, B. M. (1991). Into Postmodern Weightlessness. *Journal of Planning Education and Research* 10: 181–7.

Moos, A. (1989). The Grassroots in Action: Gays and Seniors Capture the Local State in West Hollywood, California. In *The Power of Geography*, eds. J. Wolch and M. Dear. Boston: Unwin Hyman.

Norris, C. (1990). *What's Wrong with Postmodernism*. Baltimore: John Hopkins University Press.

Norris. C. (1993). *The Truth about Postmodernism*. Oxford: Blackwell.

Olsson, G. (1980). *Birds in Egg/Eggs in Bird*. London: Pion.

O'Tuathail, G. (1992). Foreign Policy and the Hyperreal. In *Writing Worlds*, eds. T. J. Barnes and J. Duncan. New York: Routledge, 155–75.

Peet, R. and N. Thrift, eds. (1988). *The New Models in Geography*. Boston: Unwin Hyman.

Philo, C. (1991). *New Words, New Worlds: Reconceptualizing Social and Cultural Geography*. Aberystwyth: Cambrian Press.

Philo, C. (1992). Foucault's Geography. *Environment and Planning D: Society and Space* 10: 137–61.

Pickles, J. (1992). Texts, Hermeneutics and Propaganda Maps. In *Writing Worlds*, eds. T. J. Barnes and J. Duncan. New York: Routledge, 193–230.

Pile, S. (1990). Depth Hermeneutics and Critical Human Geography. *Environment and Planning D: Society and Space* 8: 211–32.

Pile, S. and G. Rose (1992). All or Nothing? Politics and Critique in the Modernism–Postmodernism Debate. *Environment and Planning D: Society and Space* 10: 123–36.

Pindar, I. (1997). Tickling the Starving. *Times Literary Supplement*, March 28: 25.

Pratt, G. (1992). Spatial Metaphors and Speaking Positions. *Environment and Planning D: Society and Space* 10: 241–4.

Pred, A. (1992). Commentary: On 'Postmodernism, Language and the Strains of Postmodernism' by Curry. *Annals of the Association of American Geographers* 82(2): 305.

Punter, J. (1988). Postmodernism. *Planning Practice and Research* 4: 22–8.

Reichert, D. (1992). On Boundaries. *Environment and Planning D: Society and Space* 10: 87–98.

Relph, E. (1987). *The Modern Urban Landscape*. Baltimore: Johns Hopkins University Press.

Relph, E. (1991). Review Essay: Post-modern Geography. *The Canadian Geographer* 35(1): 98–106.

Robins, K. (1991). Prisoners of the City: Whatever Could a Postmodern City Be? *New Formations* 15: 1–22.

Robins, K. (1996). *Into the Image: Culture and Politics in the Field of Vision.* New York: Routledge.

Sack, R. D. (1988). The Consumer's World: Place as Context. *Annals of the Association of American Geographers* 78(4): 642–64.

Sandercock, L. (1998). *Toward Cosmopolis: Planning for Multicultural Cities.* New York: Wiley.

Saunders, P. and P. Williams (1986). The New Conservatism: Some Thoughts on Recent and Future Developments in Urban Studies. *Environment and Planning D: Society and Space* 4: 393–9.

Sayer, A. (1984). *Method in Social Science: A Realist Approach.* London: Hutchinson.

Sayer, A. and R. Walker (1992). *The New Social Economy: Reworking the Division of Labor.* Oxford: Blackwell.

Schatzki, T. R. (1991). Spatial Ontology and Explanation. *Annals of the Association of American Geographers* 81(4): 650–70.

Schoenberger, E. (1988). From Fordism to Flexible Accumulation. *Environment and Planning D: Society and Space* 6: 245–62.

Scott, A. J. (1988). *Metropolis: From the Division of Labor to Urban Form.* Berkeley: University of California Press.

Scott, J. S. and P. Simpson-Housley (1989). Relativizing the Relativizers: On the Postmodern Challenge to Human Geography. *Transactions of the Institute of British Geographers* 14: 231–6.

Shields, R. (1989). Social Specialization and the Built Environment: The West Edmonton Mall. *Environment and Planning D: Society and Space* 7: 147–64.

Short, J. R. (1989). Yuppies, Yuffies and the New Urban Order. *Transactions of the Institute of British Geographers* 14: 173–88.

Simonsen, K. (1990). Planning on 'Postmodern' Conditions. *Acta Sociologica* 33: 51–62.

Slater, D. (1992a). On the Borders of Social Theory: Learning From Other Regions. *Environment and Planning D: Society and Space* 10: 307–27.

Slater, D. (1992b). Theories of Development and Politics of the Postmodern. *Development and Change* 23: 283–319.

Smith, S. J. (1989). Society, Space and Citizenship: A Human Geography for the 'New Times'? *Transactions of the Institute of British Geographers* 14: 144–56.

Soja, E. (1980). The Socio-spatial Dialectic. *Annals of the Association of American Geographers* 70: 207–25.

Soja, E. (1986). Taking Los Angeles Apart. *Environment and Planning D: Society and Space* 4: 255–72.

Sorkin, M., ed. (1992). *Variations on a Theme Park.* New York: Hill and Wang.

Storper, M. and R. Walker (1989). *The Capitalist Imperative: Territory, Technology, and Industrial Growth.* Oxford: Blackwell.

Strohmayer, U. and M. Hannah (1992). Domesticating Postmodernism. *Antipode* 24: 29–55.

Thrift, N. J. (1983). On the Determination of Social Action in Space and Time. *Environment and Planning D: Society and Space* 1: 23–58.

Thrift, N. J. (1991). For a New Regional Geography 2. *Progress in Human Geography*, 15(4): 456–65.

Thrift, N. J. (1993). For a New Regional Geography 3. *Progress in Human Geography* 17(1): 92–100.

Unwin, T. (1992). *The Place of Geography*. Harlow: Longman.

Valentine, G. (1993). Negotiating and Managing Mutiple Sexual Identities: Lesbian Time-Space Strategies. *Transactions of the Institute of British Geographers* 18: 237–48.

Ward, D. and O. Zunz, eds. (1992). *The Landscape of Modernity*. New York: Sage.

Webber, M. (1991). The Contemporary Transition. *Environment and Planning D: Society and Space* 9: 165–82.

Werlen, B. (1993). *Society, Action and Space*. New York: Routledge.

Wood, D. (1992). *The Power of Maps*. New York: Guilford.

Zukin, S. (1991). *Landscapes of Power*. Berkeley: University of California Press.

Cities . . .

2

Exploring the Postmetropolis

Edward W. Soja

As we enter the new millennium, the field of urban studies has perhaps never before been so expansive in the number of subject areas and scholarly disciplines involved with the study of cities, so permeated with challenging new ideas and approaches, so sensitive to the major political and economic events of the contemporary moment, and so theoretically and methodologically unsettled. It may indeed be both the best of times and the worst of times to be studying cities, for while there is so much that is new and challenging, there is much less agreement than ever before as to how best to make practical and theoretical sense of the new urban worlds that have been taking shape over the past three decades.

Since the emphatically urban crises of the 1960s, nearly all the world's metropolitan regions have been experiencing dramatic changes, in some cases so pronounced that what existed thirty years ago is almost unrecognizable today. Almost every urbanist agrees that this *urban restructuring* process, linked closely to broader changes in the global economy, has been particularly widespread and intense, but here agreement ends, provoking a growing interpretative bifurcation. One extreme reaction has been to proclaim that the restructuring has been so profound as to make virtually useless all traditional frameworks of urban analysis. Wholly new constructs and epistemologies are consequently called for to understand the radically transformed urban scene. At the other extreme, a much larger number of urban scholars interpret the present primarily in terms of continuities with the past, confident that *plus ça change, plus c'est la même chose*. In their view, the way we study cities is not fundamentally flawed so does not need to be radically changed, merely updated. To add to the confusion, both these extreme views are probably correct in more ways than either is willing to admit.

It is almost surely too soon to conclude with any confidence that what has been happening to cities in the late twentieth century is either a revolutionary transformation or just another minor twist on an old tale of urban life. I take a position here, however, that is closer to the former possibility than to the latter. Reflecting my own postmodern geographical praxis, I contend that there have been profound material changes taking place in what we familiarly describe as the modern metropolis, and that these changes necessitate a significant modification in the ways we understand, experience, and study cities. To highlight what has changed the most from what remains most constant and continuous, I use the term *postmetropolis* and describe the past thirty years of urban restructuring as defining a significant *postmetropolitan transition*, a selective deconstruction and still ongoing reconstitution of the modern metropolis as it has traditionally been understood and interpreted. Implied in these terms is the emergence of what can be described as *postmodern urbanism* and, along with it, new approaches to the critical study of cities and regions informed by a variety of post-prefixed modes of interpretation: poststructuralist, post-Fordist, postcolonial, post-Marxist, postmodern feminist.

In what follows, I present some brief descriptions of postmodern urbanism and the postmetropolitan transition, extracted mainly from my recently published book, *Postmetropolis: Critical Studies of Cities and Regions* (Soja 2000). These broad brush sketches present an intentionally multi-sided picture of the theoretical and political implications that can be drawn from an understanding of what has been happening to cities over the past thirty years. There are no clear and unequivocal takes, no confident and assured ways of interpreting postmodern urbanism and the postmetropolitan transition. But if I were to offer, in advance, one uncomplicated conclusion to this chapter, it would be that a critical postmodern geographical praxis offers greater possibilities for theoretical and practical understanding and progressive political practice than does any of its alternatives.

Postmodern Cities?

It is important to recognize at the outset that there are no purely postmodern cities or completely postmetropolitan urban regions, and perhaps there never will be. Continuities with the past, at the very least with distinctively modernist forms of urbanism and urbanization, persist everywhere, shaping the spatial, social, and historical dimensions of contemporary urban life. Over the past thirty years, however, every city everywhere has been experiencing, in varying degrees to be sure, a recognizable and effectual postmodernization. Just how extensive this empirical or material postmodernization has been, whether its effects are positive or negative or both, and especially how it should affect the practice of urban studies remain highly contested. But that there exists as an object of analysis something that can be described as postmodern urbanism, or, alternatively, a postmodern urban condition, is difficult to deny, as is the corollary argument that understanding this postmodern urbanism or condition is important, theoretically and practically, in making sense of the contemporary city and ongoing urban restructuring processes.

Because of their co-presence in the urban scene, it is often difficult to separate the modern from the postmodern in the specific geographies of the city. This is complicated further if one rejects the widely held idea that modernity and postmodernity are polar opposites rather than always being complexly intertwined and inseparable. Despite these complications, however, there has developed a growing literature focusing on postmodern urbanism as a distinctive condition and conditioning feature of contemporary life, especially within the broadly defined field of critical human geography. Within this specifically geographical literature there is also a complex co-presence of modernist and postmodernist interpretative perspectives and approaches, baffling those who seek greater uniformity of vision. And here too one can make similar arguments about postmodernization. There is no purely postmodern way of studying cities, but there has been a notable increase in the use and usefulness of explicitly postmodern concepts and perspectives. Particularly noteworthy have been attempts to draw together more effectively the three often antagonistic interpretative frameworks that have played such a key role in defining critical human geography: Marxist geopolitical economy, critical cultural or humanistic geography, and geographical or spatial feminism.

Understanding the postmetropolitan transition and the expanding development of postmodern urbanism thus requires a more open, flexible, and eclectic approach than has often been the case in urban geographical studies. Such understanding may foreground an assertively postmodern geographical praxis, but without losing sight of the continuing importance of critical modernist modes of interpretation and analysis. As noted earlier, drawing a clear boundary between the modern and the postmodern is neither easy nor perhaps necessary. But as this boundary-marking has become a frequent focus of attention in the critical human geography literature, especially in various extensions of the modern (Western) Marxist tradition, a brief comment on how I see the relative strengths of critical modernism and postmodernism may be useful.

Critical scholarship of all kinds involves a commitment to producing knowledge not only for its own sake but more so for its usefulness in changing the world for the better. This intentionally progressive and potentially emancipatory project has often been ascribed exclusively to modernist thought and practice, and has therefore appeared to many critical scholars to be incompatible with postmodernism. Although some of the writings of leading postmodernist and poststructuralist philosophers and theorists seem to sustain this view, I see no such categorical opposition necessarily defining the relation between modernism and postmodernism, and find it challenging and revealing to draw selectively on both in critical geographical studies of cities and regions. Modernist critical theory and modes of interpretation, including Marxism, continue to inform my explorations of the postmetropolis, especially in understanding what remains the same today as it was in the past. For example, for all my attachments to post-prefixed concepts, there are three that I do not use: post-urban, post-industrial, and post-capitalist. But as for making practical and theoretical sense of what is *new and different* in the contemporary world, and especially in urban-industrial capitalism, I turn primarily to critical postmodernism for practical and theoretical insight.

Outlining the Postmetropolitan Transition

Studies of urban restructuring over the past several decades have crystallized around at least six broad themes, each contributing to an understanding of the causes and consequences of postmodern urbanism. These themes or interpretative approaches can be seen as a sequence of three pairs, the first consisting of the major explanatory forces shaping the transformation of the modern metropolis over the past thirty years; the second focusing on the interrelated social and spatial outcomes of these new urbanization processes; and the third involving forms of reaction, regulation, and resistance arising in response to the postmetropolitan transition. Each will be described briefly, with an emphasis on their specific urban impacts.

1 Globalization of capital, labor, culture, and information flows

Perhaps the most comprehensive of these interpretative frameworks associates postmodern urbanism and the postmetropolitan transition with the compelling catch-all concept of *globalization*. Closely tied to the onset of a new Information Age and the rise of a Network Society (Castells 1996–8), the globalization discourse has become, for many scholars, an encompassing paradigm for all studies of the contemporary. It has even been described as "the successor to the debates on modernity and postmodernity in the understanding of sociocultural change and as the central thematic for social theory" (Featherstone, Lash, and Robertson 1995: 1).

Defined broadly as a spatio-temporal compression of the world and an intensified consciousness of the world as a whole, globalization has vastly expanded the scale and scope of urban life. More than ever before, what happens in cities is significantly affected by global events – and vice versa, world events are shaped by what is happening in cities no matter where they take place. This mutual interplay between the global and the local has led to the diffusion of urbanism and urbanization to every corner of the world and, at the same time, to the at least partial globalization of every urban region on earth. It has also generated new concepts such as 'glocalization' and 'deterritorialization-reterritorialization' to help understand the dynamics of this two-way flow of global-local effects and its impact on urban and regional economic development and territorial identities from the local to the global scales.

The more specific globalizations of capital, labor, and culture have had the cumulative effect of producing the most heterogeneous cities in history, and this extraordinary *diversity* (often too simply labeled multiculturalism) has become the hallmark of postmodern urbanism. Such heterogeneity is expressed in architecture and the design of the built environment, in the organization of urban labor markets, in the formation of local community and identity, in urban politics and the planning process, and in almost every facet of everyday life: eating habits, clothing styles, music, shopping, housing, voting, etc. On a larger scale, what the architecture critic Charles Jencks (1993) called the *Heteropolis* is viewed as a world city or global city, nested in a reorganized hierarchy of 'command centers' for the global economy, shaping industrial production, trade, labor migrations, financial flows, and investment decisions. Increasingly, the local and regional geopolitical economy of cities is becoming more closely tied to this far-flung global network than to national urban systems, giving rise to what some have called a 'new regionalism' as well as to such overstated claims as the end of the nation-state and the rise of a borderless world. The postmetropolis is thus first and, for many, foremost an increasingly heterogeneous global city-region.

2 Post-Fordist economic restructuring

Although almost unavoidably tied to globalization processes and new information technologies, a second discourse on the restructuring of the modern metropolis has developed with a particular emphasis on the changing relations between urbanization and industrialization. The postmetropolitan transition and its postmodern geographies are here associated primarily with a fundamental change in the geopolitical economy of urbanism emerging from the economic crises that marked the end of the long postwar boom in the 1960s and early 1970s. Especially among the geographers most closely involved in studying this crisis-generated economic restructuring process, the postmetropolitan transition is viewed primarily as an expression of a new regime of capitalist accumulation that is more flexibly organized than the rigid, hierarchical, mass production/mass consumption systems of the postwar era. This has led to connections being made between postmodern urbanism and post-Fordist geopolitical economies, from the technopoles of California's Orange County to the 'diffused urbanization' of the Third Italy. Particularly attractive to postmodern geographical analysis has been the concept of flexibility as an adaptable outgrowth of the breakdown (deconstruction) and at least partial reconstitution of established (modern) modes of life.

In the post-Fordist industrial metropolis, urban and regional geographies have been significantly reshaped through a two-sided process of deindustrialization (especially of older Fordist clusters of industrial production) and reindustrialization (the formation of new industrial spaces or districts), another of the many 'de–re' terms that echo postmodernist and poststructuralist critical theories and methods. The three sectors that have been given the most attention in the literature on the post-Fordist urban and regional economy have been high-technology manufacturing, craft-based and often design-intensive industries, and financial and business services. Each of these sectors has been involved in a complex process of both concentration and dispersal that has contributed significantly to explaining emerging patterns of uneven development within and between metropolitan areas as well as the postmodern geographies associated with the diverse impacts of globalization and new information technologies.

3 Restructuring of urban form

The combined discourses on globalization and post-Fordist economic restructuring provide a powerful conceptual framework for understanding and analysing the major forces that have been generating the new urbanization process shaping the postmetropolitan transition. The next pair of interpretative themes are concerned more with the specific outcomes and effects of these processes on the social and spatial reorganization of the modern metropolis. One of the most obvious outcomes of the spatial restructuring of urban form has been the rise of megacities, vast (post)metropolitan galaxies with population sizes reaching well beyond what was imaginable 50 years ago. Also changing dramatically has been the internal spatial structure of these globalized city-regions. In what some deem to be the final break with classic models of the orderly geography of the industrial capitalist city, such as those of the Chicago School, the sprawling postmetropolis has become much less monocentric, less focused on a singular downtown; and is no longer as easily describable in terms of distinctively urban, suburban, and non-urban ways of life, thoroughly confusing the conventional categories and basic vocabulary of urban analysis.

There have been various ways of describing this still incomplete metamorphosis of the modern metropolis: the urbanization of suburbia, the peripheralization of the urban core, the rise of Outer Cities and Edge Cities, decentralization-recentralization, the city simultaneously turned inside-out and outside-in, diffused urbanization, mass regional urbanization, the metropolis unbound, the end of the Metropolis Era. My preferred descriptive term is *exopolis*, connoting the increasing importance of exogenous forces shaping the geography of the city, the rapid growth of Outer Cities, and the notion of an ex-city, cities without the traditional traits of cityness as we have come to define them in the past. Not only has the conventional division between urbanism and suburbanism as ways of life begun to disappear, the mosaic of local worlds that comprise the traditional city has been given an almost kaleidoscopic spin.

4 Restructuring the social order

Interwoven with the jumbled spatiality of the postmetropolis is a recomposed sociality that has become similarly fluid, fragmented, decentered, and rearranged in complex new patterns. Older polarities, such as those between the bourgeoisie and proletariat, or the wealthy, the middle class, and the poor, or black versus white, persist, but a much more polymorphous and fractured social geometry has taken shape from a far-reaching restructuring of the social boundaries and categorical logics of class, income, skill, race, ethnicity, and gender that characterized the modern metropolis up to the early 1970s. In another seemingly paradoxical process, the social order has been simultaneously and selectively depolarised and repolarised, paralleling the decentering and recentering of the urban spatial order.

Much of the emphasis among those studying the restructured social order has been given to immigration, changing ethnic and gender divisions of labor, increasingly segmented labor markets, the rising power of a professional-managerial class, yuppification and gentrification of the urban scene, increasing homelessness, the growing numbers of the working poor, and the consolidation of a permanent welfare-dependent urban underclass. Summing up the impact of all these developments has been a pronounced resurgence in inequality, a widening gap between rich and poor which, in the United States at least, has reached levels last seen in the Great Depression and has become most extreme in the two largest postmetropolitan urban regions, New York and Los Angeles. Despite attempts from the Left and Right to normalize this resurgence of inequality as an expected accompaniment to capitalist economic growth, it remains the most challenging public and political finding of the literature on urban restructuring and needs to be seen as an integral part of postmodern urbanism and postmodern urban politics.

5 Carceral cities

If the previous four discussions taken together define a new 'regime' of urbanization, a distinctively postmetropolitan mode of urban development, then the next pair explore the emerging institutional, behavioral, and ideological changes that are reorganizing what can be called a postmetropolitan mode of social and spatial regulation. They help answer a fundamental question: given the exceedingly volatile postmodern geographies produced by the new urbanization processes, with their unprecedented cultural heterogeneity, widening social and economic disparities, and multiplying points of tension and confrontation based on differences in race and ethnicity, gender, income, sexual preference, lifestyle, location, and other social and spatial attributes, what has prevented the postmetropolis from exploding more frequently and more violently than it has over the past decade?

An intensification of social and spatial control has marked the postmetropolitan transition, reacting to what Mike Davis (1990, 1998) sees as the formation of an ecology of fear and the spread of a security-obsessed urbanism. The hard side of this intensification has involved improved techniques of policing, surveillance, and territorial control – what I once described as the substitution of police for polis – as well as the multiplication of many different kinds of protected and fortified spaces, from what Davis calls sadistic street environments and panopticon shopping malls to the proliferation of gated communities and the rise of privatized residential governments. In so many ways, the postmetropolis has become an archipelago of carceral cities, islands of enclosure and anticipated protection against the real and imagined dangers of daily urban life. Such fortressing of the city may reach its greatest intensity in the US, but it is spreading everywhere in the contemporary world as still another dimension of globalization (Caldeira 1999).

6 Simcities

Here the emphasis shifts to softer and more subtle forms of social and spatial regulation, the restructuring of the urban imaginary, our situated and city-centric consciousness, and how this ideological refabrication affects everyday life in the postmetropolis. At the core of this restructuring is a growing confusion between the real and the imagined, and the interjection into the blurring of what has been called hyperreality, filled with simulated images that are increasingly perceived as materially real in themselves – what Jean Baudrillard (1983) has called the "precession of simulacra." This restructuring of the urban imaginary – the fusion of the real with its representations – has been central to the conceptualization of postmodern urbanism, at least since the seminal work of Fredric Jameson (1984), and has filtered deeply into the literature on cyberspace and virtual reality, triggering such notions as digital communities and CyberCities, where hyperreality, along with postmodern urbanism as a way of life, are generated and diffused electronically.

This secular diffusion of hyperreality into everyday urban life has been linked to the theme-parking or disney-worlding of the city, the representation of urban geography as a simulated reality of enchantment, danger, cultural iconography, entertainment, diversion, globality. Just as the ecology of fear has led to the rise of carceral cities, the sprawling spaces of hyperreality have created actual Simcities, to refer to the popular computer game that bills itself as the 'original city simulator.' One no longer needs to log on to play, for an increasing portion of everyday life in the city – from shopping in themed malls and living in themed residential communities to choosing who to vote for or what political cause to join – is shaped by simulated representations of reality perceived as materially real. However one may react to this integral aspect of postmodern urbanism, it cannot be ignored.

Selected Scenes from Postmetropolitan Los Angeles

To illustrate some of the more general arguments presented above, I present here a series of indicative glimpses of the postmetropolitan transition expressed in what is perhaps the most paradigmatic postmetropolis of them all, Los Angeles. Each example is both a specific expression of the particular local circumstances and an invitation to comparative study in other places and spaces.

1 Pivot of the four quarters

I take the phrase used to describe the cosmological geometry of the ancient city and its sanctification of the four cardinal directions to illustrate a new kind of cultural quartering taking shape in the postmetropolis. Los Angeles (followed closely by the San Francisco/Bay Area and metropolitan New York) has become the largest cluster of the most demographically heterogeneous cities in the US, and probably in the world, as measured by conventional statistical indexes of diversity. In a belt of incorporated municipalities running from west to east roughly in between the downtowns of Los Angeles and Long Beach, the two largest cities in the region, this statistical diversity reaches its highest level. In Gardena and Carson, with a combined population of around 150,000, there is almost a perfect quartering, with each of the four major racial-ethnic groupings (white, black, Latino, Asian) nearly equal in size. Moreover, this near-perfect quartering has been maintained since at least the 1980 census, confounding conventional sociological theories of urban segregation.

This corridor of diversity extends eastward along the Artesia freeway to Cerritos, a municipality near the border of Orange County that was ranked as the most ethnically diverse city in the US when using an index of 14 ethnic groups (Allen and Turner 1989). In the middle of this corridor is California State University–Dominguez Hills, recently identified as having the second most diverse student body in the US, behind only Baruch College of the City University of New York. It was also recognized as one of the top ten 'cyber-universities' in the country. Whether such statistical measures of diversity translate directly into significant intercultural mixing and changes in the character of everyday urban life is still open to question. But there is growing evidence in the corridor of an explicit consciousness of diversity and of experimentation in creating new intercultural and transnational forms of identity and practice. Gardena has been widely recognized as one of the most progressive local governments in the US, an All-American City with Sister Cities in Japan, Mexico, and West Africa to reflect its diverse global ties. In 1998, the country's first Museum of Cultural Diversity opened in Carson, defining itself as "a forum for cultural collaborations through the arts."

These local developments are indicative of a wider reorganization of the residential and demographic geography of the postmetropolis. While the creation of ethnically homogeneous communities and continuing patterns of residential segregation and/or gentrification have received the greatest attention in the literature, what is perhaps most new and different in contemporary cities, and another important expression of postmodern urbanism, has been the expansion not just of multiculturalism but of what might be called intercultural fusion, hybridity, and synergy. Los Angeles was the primary locus for the formation of a composite Latino identity, comprised of Spanish and Portuguese, as well as French and English speaking populations drawn from all countries south of the US; and has also taken the lead in the development of a localized Pan-Asian consciousness around the census category Asian and Pacific Islander. These large transcultural blocs, as well as the growing connections between them, are increasingly shaping local politics and contributing to new debates about cultural politics and the nature of citizenship and civil society in the postmetropolis.

2 Off-the-edge cities

The urbanization of suburbia has been an integral part of the postmetropolitan transition. Receiving the most popular attention in discussions of this restructuring of urban form has been Joel Garreau's enthusiastic depiction of the rise of Edge Cities (Garreau 1991). Garreau describes Edge Cities as representing "the biggest change in a hundred years in how we build cities," and adds: "Every single American city that is growing, is growing in the fashion of Los Angeles, with multiple urban cores" (1991: 3). In his optimistic envisioning of "life on the new frontier," Edge Cities become the "new hearths of our civilization." In comparison to the "old downtowns," they already contain more than two-thirds of the country's office space and a growing majority of all jobs.

The growth of these Edge Cities or Outer Cities is evident in practically every major urban region in the world but, as Garreau claims, they are most numerous and diverse in Southern California. Successful Edge City development, however, has masked a much darker side of the urbanization of suburbia. Whereas the successful Edge City represents a fortuitous locational clustering of jobs, industry, housing, office space, shopping, entertainment, and transit facilities, there are other places where this confluence of resources has become cruelly out of balance. In cities such as Lancaster and Palmdale, located in the Antelope Valley in northern Los Angeles County, and Moreno Valley, 60 miles east of downtown Los Angeles in Riverside County, the urbanization of suburbia has taken a particularly pathological turn, creating what, with only some exaggeration, might be called a postsuburban slum.

The 1990 census listed Moreno Valley as the fastest growing city over 100,000 in the US. Attracted by affordable housing, mainly young and ethnically mixed lower middle-class families flocked to the area, in part also to escape the real and imagined problems of the inner city. Today, the city is described as solidly middle class, with a population of about 135,000 and a median family income close to $45,000. Beneath the appearances of comfortable suburban life, however, there have been unforeseen problems of social disruption and personal despair as intense as those seemingly left behind. With local employment growth far below what was promised by optimistic community developers, the journey to work has become an unusual burden. Many workers are forced to rise well before dawn to drive or to be taken by a fleet of vans and buses, often for more than two hours each way, to the jobs they held before moving to their affordable housing. Without a large commercial or industrial tax base, local public services are poor, schools are overcrowded, freeways are gridlocked, and family life is deeply stressed as residents contend with their location in a very different kind of Edge City.

In the high desert of northern Los Angeles County, the development of Lancaster and Palmdale was given an even greater push to excess than Moreno Valley, as the area contained both the site for a proposed international airport and a cluster of big aerospace firms associated with the sprawling Edwards Air Force Base, a major cog in the NASA network of space bases. In the 1980s, the rolling brown hills and sage-brushed sands of the Antelope Valley, pictured at the Palmdale freeway exit by David Hockney in one of his most famous photo-collages, became covered by a sea of peach and beige stucco houses with red-tile roofs selling at bargain basement prices. With the end of the Cold War and the steep decline of the region's aerospace industry, as well as the ensuing real estate crisis and economic recession of the early 1990s, the booming Antelope Valley became the site of what one observer called a "middle class implosion."

Even more so than in Moreno Valley, excessively long journeys to work have been having pathological effects on family life and personal health. Many workers spend more than five hours a day in their cars and young children are often left for more than twelve hours in daycare centers that open before dawn. Suicide rates are unusually high, domestic violence felony arrests are much greater than elsewhere in LA county, and there are more child abuse reports than anywhere else in California. Violent juvenile crime and gang membership has increased precipitously, and some shopping malls have gone as far as prohibiting entry to anyone wearing a baseball cap backward or to one side. With plummeting land values, mortgage foreclosure rates are among the highest in the US, and many of the empty homes have become filled with squatters migrating north from the inner city. Sweatshops have been set up in the area and crackhouses have been raided by police battering rams. All of this gives new meaning to the notion of the metropolis inverted or turned inside-out, and represents one of the bleakest outcomes of the postmetropolitan transition.

3 Privatopia

In reaction to the urbanization of suburbia and the so-called Third Worlding of the inner city, the upper middle class (and above) have, in Mike Davis's words (1990), reasserted their social privilege by fleeing to their own fortified cells of affluence and insular lifestyles. Reflecting this escape from the city and its attendant civic rights and responsibilities has been the phenomenal growth of what are called CIDs (Common Interest Developments), in which residents own or control common areas and shared amenities, and are bound by contractual agreements enforced by a private governing body, or community association. Some lawyers have called these contract-bound housing estates "association-administered servitude regimes." The number of CIDs in the US is probably approaching 100,000 and they have become the principal form of new home ownership in almost every metropolitan region. Many CIDs are themed (and contractually maintained) around a chosen image for the community: New England Village, Greek Island Villa, Golfer's Paradise, Leisure World. Overlapping with CIDs but also present in poorer communities (although usually with some form of community association) are gated communities, which today number around 30,000, with nearly 10 million residents.

The densest concentration of CIDs, themed housing developments, gated communities, and active homeowners associations (HOAs) are probably found, not unexpectedly, in Southern California. Three distinct swarms of gated communities are identifiable: in the Palos Verdes peninsula near the Port of Los Angeles, where several whole municipalities are gated, including Rolling Hills, reputedly the wealthiest community in the US with an average household income of more than $300,000; in coastal and southern Orange County, perhaps the largest Outer City in the US and home to the archetypal retirement community at Leisure World (actually, there are two Leisure Worlds, one more upscale than the other); and, more recently, in the West San Fernando Valley and adjacent Ventura County, site of a burgeoning new technopolis and represented most tellingly by the city of Hidden Hills, where white picket fences are virtually, if not contractually, mandatory. Also located in this third swarm is a place that calls itself "the Island," centered in a 160-acre man-made lake and bristling with video surveillance cameras, electronic listening devices, roving security patrols, and other accoutrements of security-obsessed postsuburbia.

Usually not so visibly fortressed and isolated, and more engaged in the public arena than the more privatized CIDs, the many dozens of homeowners' associations (HOAs) have been particularly active in shaping urban and regional development in Los Angeles. Advocating controlled growth and greater local autonomy, as well as more self-serving objectives and privileges, HOAs have played a political role that is ambiguous and difficult to characterize, at times supporting progressive causes in alliance with low-income communities but most often simply defending their affluence in what Davis termed suburban separatism. Such separatism is exemplified by current efforts, led by HOAs, to break off the San Fernando Valley from the City of Los Angeles and establish an autonomous municipality of well over a million inhabitants. If successful, it would be the first such territorial split of an established major city in US history.

It is estimated that there are now more than 200,000 Residential Community Associations (RCAs) of various types across the US, ranging from the poorest neighborhoods to the richest island fortresses. The overall effect of this extraordinary collective privatization of urban residential space has been interpreted in many ways. Some see it as a destructive erosion of public space and civic democracy, a "secession of the successful" that is contributing significantly to the resurgence of urban inequalities and polarization. Others see it as brewing a major transformation in urban (and national) politics, with the RCA becoming the residential version of the private corporation as a form of collective property ownership and banding together to create a powerful new political force. Still others see the empowerment of residential communities as potentially contributing, through coalitions of like-minded associations and NGOs, to a reassertion of interest in local and environmental issues and more forceful movements to serve the needs of the working poor. It is all these, I believe, and more. Like so much of postmodern urbanism, it can be blindly celebrated, summarily condemned, or blithely ignored, but none of these choices add anything to our critical understanding and postmodern geographical praxis.

4 The boiler room

The diffusion of hyperreality into everyday life has added a new layer to the postmetropolitan landscape, an intrusive geography of make-believe that is forcing its way in between the materially real and concrete world of spatial practices and the representational world of the imagination. Where this layer is thickest, it engenders specialized Simspaces and Simplaces of various kinds, and these spaces and places of hyperreality are as much a part of postmodern urbanism as anything else. Disneyland in Orange County is perhaps the ancient iconic birthplace of this intrusive geography, but over the past thirty years the contextual blurring of the real and the imagined has expanded significantly in scale and scope, creating new kinds of playgrounds for what I once described as the "habitactics of make-believe." One such iconic place is the 'boiler room,' a kind of sweatshop of the New Information Age and a productive node in the manufacturing of hyperreality.

Boiler rooms are found practically everywhere today, but they probably reach their peak density in Orange County, nationally recognized as the fraud capital of America. Named for the intensity of activity crammed into the barest of spaces, the typical boiler room is a telemarketing business serving to collect money for charities, public and private institutions, credit card and loan applications, and other ventures, including occasionally fraudulent investment schemes. The basic work is done by young men and women, often starting out their employment careers, on high-tech noise-controlled phones in tiny cubicles earnestly reading from scripts prepared for the particular purpose. Some Orange County boiler rooms have a gross take of $3 million a month, and many are, directly or indirectly, involved in what is legally defined as fraud, with victims losing from $40 to $50,000 according to local authorities. During one police raid on an adventurous telemarketing company, a placard was found that captures the duplicitous honesties of the habitactics of make-believe that characterize the boiler room. It proudly proclaimed: "We cheat the other guy and pass the savings on to you!"

The boiler room is just one representative place in what is a much more extensive 'scamscape,' where the difference between reality and fantasy, fact and fiction, legal and illegal, objectivity and subjectivity,

honesty and fraud is sublimated behind the invasive power of spin-
doctored representations and simulations. The particularly fulsome
scamscape of Orange County has been the site of some of the worst
defense industry frauds (e.g., falsifying tests on missile equipment
believed by the makers to be of unquestionable quality) and was deeply
involved in the vast Savings and Loan scandal in the late 1980s, which
cost US taxpayers upwards of half a trillion dollars to cover up and
repair. In late 1994, the scamscape exploded again, when the entire
Orange County government declared bankruptcy after it was found that
the tax collector ran the county's finances very much like the Savings
and Loan charlatans, with a touch of the boiler room thrown in for
good measure, gambling the county treasury on inverse floaters, reverse
repos, and other magical financial attractions of the cyberspatial stock
and bond markets (Soja 1996: 274–8). Here then was a game of Sim-
county being played by a Simgovernment on behalf of Simcitizens with
an uncomfortably real impact.

5 Florence and Normandie

I refer here to the most globally visible site of the violent urban upris-
ing that took place in Los Angeles in late April–early May of 1992,
locally referred to as the Justice Riots. What occurred on the corner of
Florence and Normandie was not just a repeat of what happened during
the Watts Rebellion of 1965, but reflected, in addition to continuing
racism and police violence against African Americans, the beginnings of
a new order of postmetropolitan urban life. Seen in retrospect, it was
an event that symbolized in not so easily perceived ways a significant
turning point in the postmetropolitan transition, a moment when
thirty years of crisis-generated restructuring began to break down in
what can be described as a *restructuring-generated crisis*, a crisis
arising from the very nature of the postmetropolis and postmodern
urbanism.

As the violent attack on white truck driver Reginald Denny by black
youths, enraged at the acquittal of the police officers responsible for the
filmed beating of Rodney King, was beamed around the world, a pecu-
liar revisioning of power and governmentality was also happening in
Los Angeles. Established modes of social and spatial control in the
Carceral cities and Simcities were themselves being attacked and visibly
disrupted. By failing to confine the riots and looting to their proper
places, the police failed also in the surveillance and control of the pur-
ported 'enemy within,' sending chilling signals to even the most heavily
guarded and gated communities. One effect of this breakdown in

control among those who felt most threatened by the readily apparent expression of angry black power was a resurgence of well-armed and vigilant white militias seeking independent forms of defense and survival in the postmetropolis, a backlash primarily of frustrated angry white men that has exploded throughout the US many times in the ensuing years. More subtle in its manifestations was the immediate reaction of the black youth at the Florence and Normandie crossroads (and beyond).

Captured in the videotaping of the Reginald Denny beating was the first act of what can be seen as a documentary theater performance of an insurgent urban imaginary, deeply knowledgeable of the workings of the media and the pervasive power of simulations and spin-doctoring. Staged on that fateful corner were two happenings, one a brutal beating in frustrated retaliation to the Rodney King verdict, the other a subversive minidrama about the material and symbolic hyperrealities of racism in SimAmerica. The second performance began with a pointed question: if a videotape of many white men kicking and bludgeoning a lone black man could be dismissed as not being what it appeared to be, would it be possible for the same result to occur with a videotape of many black men equally brutally attacking a lone white man? Its denouement, recalling in a most contemporary way Gunnar Olsson's allusions (in chapter 12, above) to Magritte's painting *Ceci n'est pas une pipe*, was played out at the trial of the 'LA Four' charged with the attempted murder of Denny.

At the trial, the mother of one of the defendants was shown some stills from the videotape. They portrayed, full-face on, staring straight at the helicopter camera with fist raised high, what appeared very clearly to be the mother's son Damian. Asked if she could identify the person in the picture the mother was firmly noncommittal, saying that, well . . . , hmmm . . . , it could be my son, but then again, you know, pictures can be very deceiving, I really don't know for sure. Through it all, Denny refused to vilify his attackers, instead referring with antiracist gratitude to the four nonwhite citizens who came to his rescue. Eventually, nearly all the charges against the LA. Four were dismissed except in the case of Damian Williams, the lead performer, who was sentenced to a maximum of ten years for his subversive actions. The important issue was not the result of the trial itself, however, but what it brought to the surface: the potential power of a strategic response to the oppressions of race, class, and gender that uses the electronic media to project a new kind of 'iconic radicalism' (see Olalquiaga 1992) aimed at subverting and transforming the powerful images shaping contemporary urban life. Such battles to shape prevailing images and the urban imaginary for progressive purposes, regardless of what may be materially or

legally factual, are likely to play an increasing role in the politics of the
postmetropolis in the future.

6 The Bus Riders Union

The reconstituted Inner City of Los Angeles, now in some areas as
densely populated as New York, has become filled with a majority of
minorities, perhaps the largest concentration of the immigrant working
poor and the 'truly disadvantaged' in the US. While prevailing images
of such places as South Central paint only a picture of violence, crime,
and despair, the actual Inner City has (also) become an extraordinary
site of recovery and renewal. Especially noteworthy have been post-
1992 coalitions and solidarities that have used the specific geography of
the postmetropolis as a strategic staging point for successful political
struggles for *spatial justice* and *regional democracy*. These struggles
began earlier as vigorous movements for specifically environmental
justice, but have recently expanded to a more general response to
postmodern urbanism as an inequality-generating and polarizing spatial
system. Perhaps the most successful of these strategically spatial
ventures has been the Bus Riders Union (BRU), a broad coalition of
the working poor that challenged the spending priorities of the
powerful Los Angeles Metropolitan Transit Authority in a class
action suit on behalf of 350,000 'transit-dependent' bus riders . . . and
won.

The BRU and its lead counsel, the NAACP Legal Defense Fund,
linked civil rights legislation to the geography of transit use in Los
Angeles to argue that transit-dependent bus-riders (primarily the immigrant poor) were being discriminated against by the policies and investment patterns of the MTA, which clearly favored predominantly
wealthy, white, suburban users of the expensive new fixed-rail system
being constructed. Even among all bus riders, it was shown that the
wealthy were receiving greater subsidies than the poor. The discrimina-
tion was therefore not only racial, it was spatial as well. Led by the
Labor/Community Strategy Center (LCSC), an activist organization that
developed out of the struggles to prevent factory closures in the 1980s
and a pioneer in the local environmental justice movement, the BRU
succeeded in 1996 in winning a Consent Decree that would, if fully
implemented, induce a massive reallocation of public funds – "Billions
for Buses" as the new program is described – to serve the transit needs
of the poor, minority, immigrant, and largely female Inner City population. This stimulated Eric Mann, one of the leading figures in the
LCSC, to write a book on the case, ambitiously titled, *Driving the Bus*

of History: The LA Bus Riders Union Models a New Theory of Urban Insurgency in the Age of Transnational Capitalism.

At the very least, the BRU case represents an important example of coalition-based movements arising around the spatially specific intersections of race, class, and gender in the postmetropolis, and aiming to stimulate a new sense of spatial justice and regional democracy, of citizenship defined in part around rights to the city, the residential and associative rights – and responsibilities – of all urban dwellers to participate and shape the social production of their lived spaces. A similar spatial logic and consciousness has characterized more recently developed coalitions such as the Los Angeles Alliance for a New Economy, which has taken the lead in struggles over a Living Wage and become part of an even larger movement that has helped make Los Angeles a major center of new labor initiatives in the US.

Once described as a 'nonplace urban realm,' filled with 'communities without propinquity' sprawled out without local neighborhood ties, and today recognized as the country's densest cluster of privatized communities and lifestyles, Los Angeles has also become an exemplary center for progressive place-based and spatial politics, for a reassertion of the local in an age of globalization, and for a strategic postmodern geographical praxis.

REFERENCES

Allen, J. P. and E. Turner (1989). The most ethnically diverse urban places in the United States. *Urban Geography* 19: 523–39.

Baudrillard, J. (1983). *Simulations.* New York: Semiotext(e).

Castells, M. (1996–8). *The Information Age: Economy, Society and Culture,* 3 vols. Oxford: Blackwell.

Caldeira, T. P. R. (1999). *City of Walls: Crime, Segregation, and Citizenship in Sao Paulo.* Berkeley and Los Angeles: University of California Press.

Davis, M. (1990). *City of Quartz: Excavating the Future in Los Angeles.* London and New York: Verso.

Davis, M. (1998). *Ecology of Fear: Los Angeles and the Imagination of Disaster.* New York: Metropolitan Books–Henry Holt.

Featherstone, M., S. Lash, and R. Robertson, eds. (1995). *Global Modernities.* London: Sage.

Garreau, J. (1991). *Edge City: Life on the New Frontier.* New York: Doubleday.

Jameson, F. (1984). Postmodernism, or, the cultural logic of late capitalism. *New Left Review* 146: 53–92.

Jencks, C. (1993). *Heteropolis: Los Angeles, the Riots, and the Strange Beauty of Hetero-Architecture.* London: Academy Editions.

Olalquiaga, C. (1992). *Megalopolis*. Minneapolis: University of Minnesota Press.

Soja, E. W. (1996). *Thirdspace: Journeys to Los Angeles and Other Real-and-Imagined Places*. Oxford: Blackwell.

Soja, E. W. (2000). *Postmetropolis: Critical Studies of Cities and Regions*. Oxford: Blackwell.

3

Postmodern Geographical Praxis? The Postmodern Impulse and the War against Homeless People in the 'Post-justice' City

Don Mitchell

I cannot help but be impressed at the way in which a whole world of thought and cultural practice, of economy and institutions, of politics and ways of relating, began to crumble as we watched the dust explode upwards and the walls of Pruitt-Igoe come crashing down.
David Harvey 1989b: 258–9

A [postmodern] planning theory must be built upon an epistemological openness and flexibility that are suspicious of any attempt to formalize a single, totalizing, way of knowing, no matter how progressive it may appear to be.... This means not only tolerating difference but encouraging what can be described as *the disordering of difference* (as opposed to the modernist search for order and stability).
Edward Soja 1997: 245–6

What would a postmodernism from below reveal?
Michael Dear and Steven Flusty 1998: 53

Pruitt-Igoe and the Politics of the 'Postmodern' City

Perhaps it is by now a cliché, but there just might be something to dating the beginning of postmodernism to that moment – 3:32 p.m. on July 15, 1972 – when the Pruitt-Igoe public housing project was dynamited. For Charles Jencks (1981) and after him (and more critically) David Harvey (1989a, 1989b), this moment represented the final acceptance of the failure of high-modernism as an architectural move-ment, but also as a way of conceptualizing urban social relations. The Corbusian vision of a gleaming city of 'machines-for-living-in' had yielded instead a brutal and brutalizing urban environment, as cold and (ironically) inefficient as it was universalist. Now architecture, and with it urbanism as a whole, would 'learn from the local' and vernacular, seeking to recreate a form of urbanism built on the small scale, the neighborhood, the contextual. Jane Jacobs' (1961) organic, and rather nostalgic, vision of the street – her attack on modernist planning – became the mainstream, at least on the drafting tables in the studios of leading design schools. It is, of course, the template for the 'new urban-ism' (Kunstler 1993; McCann 1994; Till 1994), and for the gentrifica-tion of countless warehouse districts, row-house neighborhoods, and even commercial strips (for contrasting discussions, see Ley 1996; Smith 1996). Rather than focusing on the sorts of universal needs (e.g. for shelter) that gave rise to public housing projects throughout the 'devel-oped' world in the middle decades of the twentieth century, many urban theorists and policy-makers now dilate on the means to promote local and contextual 'difference,' even to the degree that it might become 'dis-ordering' (which is oddly seen as a good in and of itself).

If the implosion at St. Louis represented a radical transformation of urban form and social relations, then it also seemed to indicate, as David Harvey remarks in the quotation above, a break in the way that social life and thought more generally were organized. The representation of this break in academic discourse is now at least as clichéd as the invo-cation of Pruitt-Igoe: where once the goal of social life was the 'modern' promotion of order and control, now the 'postmodern' world is gov-erned by 'anarchy' and 'chance'; where once there had been a 'modern' dream of 'permanence,' the 'postmodern' thrives on 'transience'; in the place of 'narrative' has come 'discourse'; instead of 'the real,' we live amid the 'hyperreal'; rather than being 'rooted,' our identities and rela-tions are 'rhyzomatic.' The 'modern' realm of necessity has been replaced by the 'postmodern' realm of desire, purpose by play, depth by surface. Hierarchies have given way to grids and networks, strategic planning to contextual planning, massification to demassification, Tay-lorism to flexible specialization, Fordism to "just-in-time."[1] The desire for similarity, the sort of similarity that assumed that a single large-scale, universalist design for public housing would efficiently create a new kind of citizen even as it tempered the excesses of the capitalist housing market, seems to have been replaced by a new-found respect for 'dif-ference.' Or so the cliché goes.

In reality, the nature of the 'postmodern' city (or even its existence) is a highly contentious issue, both within academic discourse and within the contemporary city itself. Analysts and activists, as well as politicians and pundits, square off over the meaning of urban space, the nature of social control within the new city, and the degree to which the 'city of difference' is only such because it is also (or should be) the city of exclusion (Davis 1991; Sorkin 1992). Behind these debates is the even more contentious question of agency, culpability, and power. Is the 'postmodern' city an achievement or an accident? For Michael Dear and Steven Flusty (1998: 63), the postmodern city is a city largely devoid, in any overarching way, of assignable culpability: rather the urban landscape is accidental. It is "not unlike that formed by a keno gamecard" in which outcomes are the result of "some random draw."[2] For Dear and Flusty, this new landscape is evidence of a "radical break in the way cities are developing" (1998: 50), and their goal, following Derrida, is to "rehearse the break" – to assume that a radical break in urban practices has occurred so as to release "our capacity to recognize it" (1998: 50). A postmodern geographical praxis,[3] by implication, is therefore one that provides "a new way of understanding cities" (1998: 68) so as to better intervene in new urban practices.[4] Such an intervention, as Edward Soja (1997) argues, should take the form of a project of "planning in and for postmodernity." Postmodern planning begins from the assumption of a radically complex and anarchic social world and requires that we encourage the *further* disordering that difference and a radically overdetermined political economy alike bring, that we promote the development of "Thirdspaces" (Soja 1996) for radical social action, and that we "choose the margins" (1996: 319) as the locus of progressive political practice.

For David Harvey (1989a), by contrast, beyond the "surface froth" that has caught the eye of postmodern theorists, there is little random about urban transformation in the period after Pruitt-Igoe. Political action, must therefore be *centered* on radical political-economic change. Careful attention to structured crises within capitalism – and particularly the means by which those crises are postponed or transcended – therefore helps us to understand how urban space is structured and how it is understood by those who use it, those who own it, and those who seek to control it. Both modernity and postmodernity, Harvey (1989a: 117) argues, are marked by "fragmentation, ephemerality, and chaotic flux," as well as by what he has since come to call "structured permanances" (Harvey 1996).[5] The trick, in both the 'modern' and the 'postmodern' worlds, is to understand how these fluxes and permanances come together in specific places and at specific times. Yet Harvey (1989a) does argue that there has been a marked 'sea change' in the way social life is structured and made known, a change that has gained increasing speed and importance in the years since Pruitt-Igoe.

Following Marx, Harvey argues that a quantitative change – in the rate of capital circulation, for example – has given rise to a qualitative shift in social structures, ways of knowing, and cultural practices. For Harvey, a postmodern geographical praxis would be one that seeks to understand and reveal the social structures behind these quantitative and qualitative changes so as to find ways to take control of them, to socialize them, and to democratize them. For this "metatheory cannot be dispensed with" (Harvey 1989a: 117).

Despite their differences, then, Dear and Flusty, Soja, and Harvey all argue that the processes structuring the urban landscape – and social relations within that landscape – have been radically transformed in the years since Pruitt-Igoe was imploded, but where Dear and Flusty see an "apparently random" set of processes operating beyond the realm of the landscape itself, where they see a city structured through anarchic forces, Harvey seeks to show the deep structural tendencies that give rise to this "surface appearance." Between these, Soja sees a "Third-space" of liberatory social possibility.[6] This debate is important, because each position leads to a different concept of social justice and different proposed means of struggling towards that justice. And here, that implosion in St. Louis takes on new meaning.

Postmodernism and social justice

If the demolition of Pruitt-Igoe stands as a ready symbol of the change from the modern to the postmodern sensibility in architecture, in the construction of the urban landscape, in the nature of social relations, and in the discourses of the academy, then it also stands for something else: with only a little exaggeration, it is also one symbol among many of the demolition of any remaining faith in universal human emancipation, universalizable justice. Since the early 1970s – or perhaps more accurately since the defeat of the Paris and Prague Springs of 1968 – the loss of faith in universal human emancipation and universalizable justice is only heralded or mourned (depending on political position), and rarely struggled for, rarely seen as a living possibility (see Singer 1999). Further advanced by the crushing of the Beijing student movement in 1989 and the rapid capitalist cooptation of the revolutions in Eastern Europe that same year, this transformation of the very models through which we understand the world – from at least some faith in the *possibility* of universal human emancipation to a celebration (or perhaps resigned acceptance) of 'the local,' the differentiated, and the situational as the only acceptable yardstick of justice – has had profound effects on the way lives are led in contemporary cities. It has had just as profound an effect on the way that oppositional social struggle is theorized. A corollary of the belief that capitalism is now 'disorganized' (Lash and Urry 1987; see also Gibson-Graham 1996), and that urban outcomes are random rather than structured, the localization of social justice is rationalized on the argument that the only way to combat such a system is through localized struggle responding to the exigencies of the moment. Gayatri Spivak, for example, argues that "my politics are only directionless to the extent that global capitalism is directionless. In this situation, when there is a constantly active and impersonal loop of finance capital, the only way that resistance works is through critique on small fronts that are immediately global. You have to constantly foil the localized efforts – patenting DNA, population control, pharmaceutical dumping, etc." (quoted in Wallace 1999: 20). The implication is that the struggle for justice must itself become localized so as to combat the localizing effects of global capitalism: there can be no systemic critique operating at the scale of capital itself. Besides being a fully defeatist position, this is really little more than bad dialectics: confusing levels of abstraction with the scale of processes (cf. Cox and Mair 1989).[7]

If one turns one's attention from its architecture and towards the social relations – and political practices – it embodied, Pruitt-Igoe can be seen as an important symbol of this transformation from the struggle for universalizable justice, from critique at the scale of the system, to a faith in the local. For what was destroyed in St. Louis was not just a modernist building, but *state-subsidized housing*. It was not just the dream of a sleek modernist future that was dynamited, but also the very idea – in America at least – that society as a whole has a collective obligation to improve the environment in which we, all of us, live. It marked the beginning of the end of a societal, *collective*, obligation to assure that housing is decent and affordable. Pruitt-Igoe was a symbol, however flawed,[8] of a social commitment to housing the poorest of urban residents, to making the city at least partially *theirs*. That was a commitment to social welfare that had been won – and only grudgingly given – through concerted social struggle.[9] Pruitt-Igoe, perhaps unintentionally, quite clearly marks the end of that social commitment, just as it provides a strong symbolic image of the end of any faith in the collective project of progress and emancipation. These two symbolic aspects of Pruitt-Igoe are not at all disconnected. Rather, the post-Pruitt-Igoe, post*modernist* city, is, quite clearly, the privatizing and atomizing city: neoliberalism and postmodernism go hand in hand (Katz 1998a). If the postmodern city is the city of surfaces, play, and the fulfillment of desire, of the local and the contextual, then it is also a city of deprivation and want (two words that never appear in all the lists of the characteristics of postmodernity).[10] And it is a city whose well-off residents have little interest in doing anything substantive about that deprivation and want. The timing of Pruitt-Igoe's destruction was not accidental: it was a symbolically important cog in the Nixon administration's 'new federalism' that began, long before President Clinton could finally kill it off, the dismantling of any national commitment to social, *public* welfare (Piven and Cloward 1993: 344–406).

Indeed, from Nixon's first version of a new federalism that devolved housing funds (and authority) to the various state governments (and instituted the first plans for housing vouchers) to the Clinton administration's current policy allowing public housing units to be destroyed even when no new ones are constructed to replace them, postmodern urbanism has been, above all else, a *privatized* urbanism. Whole realms of social reproduction, as Cindi Katz so importantly notes (see chapter 4 in this volume; also, 1998a, 1998b), have been auctioned off wholesale to private police forces, prison operators, parks conservancies, business improvement districts and the like (see also Kodras 1997). Any idea that there is a *public* good, a *common* weal, has been pushed aside, as the remainder of this paper will show, for a punitive form of urbanism meant only to protect the interests of the privileged few. *This*, I suggest, *is* the postmodern city, because it is the form the city has assumed *after* the end of the progressive modernist project of universal, common, public, *collective* emancipation. The development of such a city is only advanced by a retreat by critical, presumably progressive social theorists from the desire to understand it in its complex, *structured*, totality – in the retreat from 'metatheory' into a theoretical world that assumes that because something appears, from one perspective, to be 'random,' it must be so. To the degree I am correct, postmodern geographical praxis, as it unfolds in the academy, has become little more than a handmaiden to the brutal postmodern geographical praxis of the streets – despite all intentions to the contrary.

The Postmodern City: Heterotopias of the Homeless?[11]

But to make such an argument, we must turn away from geographic debates over postmodernism and social justice and look instead at what 'postmodernism' as seen 'from below' reveals (to appropriate the formulation of Dear and Flusty). What does "new planning" (to use Soja's phrase) in the postmodern city look like? The vision is not salutary. Pruitt-Igoe was never adequate to the needs of its residents, but it at least represented some form of commitment on the part of the state and the population as a whole – however contested – to assuring that even the poor had a *right* to a place to sleep, a place to call home.[12] That commitment, and hence that right, no longer exist in America, not even in rhetoric. At any significant level of discourse, such a commitment is not even recognized as valuable. Instead, cities, with the active support of the federal government, using as an excuse some presumed 'compassion fatigue' among the populace, have created a form of urbanism based on penalizing and criminalizing those who have no homes, no places to stay, those who, as one law scholar puts it, are "the problem lying on our sidewalk" (Paisner 1994).[13] In the place of a commitment to social housing has come a legal war on homeless people (Mitchell 1997b). Where once the way to solve the problem of homelessness was to struggle for better housing, better services, and for increased political power for poor people (Piven and Cloward 1977), the way to solve the problem of homelessness in the postmodern city – the city *after* the modernist Pruitt-Igoe was demolished – is not at all to reinvigorate the public housing program, nor to struggle for the funding of effective and humane means for intervening in the lives of the mentally ill (see Winerip 1999), the drug addicted, or the alcoholic, but through punitive laws that make it impossible for homeless and other street people simply to *live* (at least without breaking any laws) (Waldron 1991; Foscarinis 1996; Mitchell 1997b, 1998a 1998b). If Pruitt-Igoe represents the end of modernism, then it also represents the end of that aspect of modernism that saw the city as the site for constructing a more *just* society. It represents the end of an ideal in which the people of the city were entitled to – had the right to – minimally decent housing, in which people had access, through state intervention, to education and jobs, and in which the steady advancement of decent lives for as many as possible was a primary goal of political and social struggle.[14] Instead, political activists and critical theorists are directed, in Edward Soja's (1996) neologism, to search out some "Thirdspace" for "resistance and transgression" (320), a "space of extraordinary openness," where we can learn to engage in "thirding-as-Othering" (5) and where we can learn to move beyond "all binarisms" (5) – presumably including 'housed/not housed.'

For many postmodern urban theorists, this "Thirdspace" aligns with the idea of heterotopia as described by Foucault (1970, 1986). Foucault's (1986) own description of 'heterotopia' is as confused as it is confusing, and not particularly helpful as a guide for social practice – especially social practice that has as its goal social justice. It is hard to imagine what sort of practices should be devised in the struggle to locate or create spaces that are at once "a space of illusion that exposes every real space, all the sites inside of which human life is partitioned, as still more illusory" and "a space that is other, another real space, as perfect, as meticulous, as well arranged as ours is messy, ill constructed and jumbled" (1986: 27). Nor is it easy to understand what Foucault has in mind when, as examples, he points to everything from hotel rooms to prisons, from Roman baths to insane asylums. This has not deterred many postmodern theorists from latching onto the notion as a suggestion for the sorts of spaces we should be struggling toward, struggling to protect and maintain, struggling to support as the key sights for the production of a liberatory 'difference' and the locus of 'resistance.'[15] This is so because contemporary heterotopias (like psychiatric hospitals and prisons) are, as Soja (1996: 159) notes, spaces of 'deviance,' and thus sites for the production of just that 'difference' authors like Soja want to promote: mental illness is held up as a site for liberation, not something to be liberated from. Beyond that, as even Soja (1996: 162) admits, "Foucault's heterotopologies are frustratingly incomplete, inconsistent, incoherent." Even so, he argues that "they are also marvellous incunabula of another fruitful journey into Thirdspace, into the spaces that difference makes, into the geohistories of otherness" (Soja 1996: 162). Heterotopias, whatever they may be, are key loci of social change.

For the homeless, however, no matter how marvelous such a journey promises to be, no matter how extraordinarily open is Thirdspace, that Thirdspace, that heterotopia, would almost certainly be exchanged for the more confining space of a home with walls, a roof, a private toilet, and a bed to call one's own. Prison or the asylum might be a radical heterotopia of resistance, but it is hardly a substitute for a decent social housing program. For the postmodern American city – the city after Pruitt-Igoe, but also after "the end of welfare as we know it" (to use the most telling phrase in the 1990s liberal lexicon), the city after the rise and now the resigned acceptance of permanent homelessness – is in desperate need not of more prisons, but more affordable housing. Some 750,000 people are homeless in the United States every night; around 2 million are homeless during the course of a year.[16] Since the early 1980s, funding for federal housing programs in the United States has been cut by 75 percent. About 3 million poor people pay more than half their income in rent alone; by 1993, 5.3 million households either paid more than half their income on housing, lived in physically dilapidated structures, or both. In the eight years after 1984, the number of affordable housing units in cities fell by 478,000.

Homelessness in America is a permanent crisis (though certainly not an unsolvable one); the treatment – or really the lack of it – for the mentally ill is utterly shameful (see Winerip 1999). Jennifer Wolch and Michael Dear (1996) have argued quite forcefully that such a state of affairs is a product of "malign neglect." But they do not go far enough, and they misplace the nature of the neglect. For on the city streets it is not neglect at work, but an active pogrom, an active seeking for a solution to the problem of homelessness that attacks not the processes and social relations that produce homelessness, but homeless people. Homeless people are not at all neglected by those who would do away with them; quite the opposite. The neglect, instead, resides (with the important exception of Michael Dear and Jennifer Wolch)[17] in the 'postmodern' academy as it searches out new examples of local transgression, new evidence of the radical openness of "Thirdspace." And this neglect is particularly malign since the pogrom directed towards homeless people is not incidental to the construction of the postmodern city; it is integral to it. It is time we turned our attention to postmodern geographical praxis as it has developed in the last three decades on the city streets of America. I will do so by focusing on two specific moments in this war against homeless people: their demonization as "broken windows" (and thus worthy targets of the law), and the subsequent proposed move, through zoning and planning, to remove the most intractable of the homeless from the spaces of the 'city of difference.' And, I will show, this latter move is not at all inconsistent with the main tenets of 'postmodern' urban theory and planning, at least as it has been represented in geography. It aligns quite readily with what I see as the making, after Pruitt-Igoe, not of some liberatory post*modernist* city, but a fully repressive post-*justice* city.

The War on Homeless People, Round 1: Broken Windows

Antihomelessness

It's hardly news anymore to point out that we in the United States are in the midst of a war not on poverty, but on poor people – a war that has been enormously popular with middle-class voters. New York's mayor Rudolph Giuliani rose to political prominence on his plans to clean the streets of homeless people, beggars, 'squeegee men,' and street peddlers (though he suffered something of a setback when he targeted urban gardeners) (cf. Smith 1998). These highly publicized 'quality of life' campaigns have been lauded by some, like the influential conservative thinktank, the Manhattan Institute, as the very thing that has "saved" the city. Crime rates are down, people are spending more in city stores and restaurants, real estate prices are up (way up), and there is, as the *New York Times* frequently boasts, a new buoyancy in the city. In San Francisco, Mayor Frank Jordan pursued a similar agenda, and was so successful, at least rhetorically, that his quintessentially liberal successor, Willie Brown, has retained the most regressive components of his anti-street people campaign. In almost all cities of any size, including many quite small ones, mayors and city councils have loudly proclaimed their desire to "take back the streets" from precisely those who have been thrown onto them: the refugees from the war on social welfare (and they have passed countless regressive laws designed to do just that: see NLCHP 1991, 1993, 1994, 1996).

The argument behind these campaigns is quite simple: in order for a decent city life to be possible; in order for downtowns to remain vibrant and interesting; in order for capital to find urban rather than suburban settings the place to be; in order, as we shall see, for there to be the sort of diversity and difference postmodern urbanism hopes to promote; in order for all these things to be possible, public order must be maintained in public space (MacDonald 1995; Simon 1992). Public order is important because it is a means of making space attractive to tourists, housed residents, suburban shoppers and commuters, and capital in general (see Kelling and Coles 1996; Tier 1993). And the way to promote order is to promote discipline by policing against sleeping in public, urinating in alleys, aggressive panhandling, or otherwise making a 'nuisance' of one's self, without regard to whether such policing prohibits homeless people from doing what they must in order to survive.[18] According to the Manhattan Institute's Heather MacDonald (1995: 80), evidence from San Francisco suggests that "merely enforcing longstanding norms of public conduct may have far more effect on reducing disorder than any number of social programs."

And that's the crux of it. The elimination of social welfare, including the right to housing, is not only possible, but desirable: it will make for a better city. After all, MacDonald (1995: 80) goes on to argue, providing housing and other services to the homeless is "unavailing" unless accompanied by strict regulation of homeless people's behavior, including control over their discretionary income, itself an expensive proposition. Better to find a way to discipline homeless people without the expense of actually housing them. The stringent policing of public space is seen by many proponents of antihomeless laws as a far more efficient means of producing a 'liveable city' (for some!) than providing decent housing, the wherewithal to live, or the social – and economic – intervention necessary to humanely assist those who are mentally disturbed.[19] Hence the sorts of laws that are now commonplace in United States cities: in Dallas it is illegal to "sleep or doze in a public place"; in San Francisco, it is illegal to "camp" anywhere in the city; in Reno, Nevada, one may not remain in a park for more than four hours at a time; in Atlanta, it is illegal to walk across or stand in a parking lot if you do not have a car parked there; in Seattle it is against the law to sit on sidewalks or curbs; in New York it is a crime to sleep on trains; in Chicago it is illegal to beg; in almost all cities "aggressive" panhandling is illegal (NLCHP 1994, 1996). Couple all that with frequent city 'sweeps' of homeless people (cf. Simon 1992; 1995); with stepped up campaigns against public drunkenness or urinating and defecating in public (San Francisco's "Matrix" program is perhaps the most prominent example); and with campaigns to close or remove homeless shelters from key urban locations (Dear and Wolch 1987; Mair 1986; Wolch and Dear 1996); and what we have is, simply and starkly, a war against homeless people – an attempt to cleanse them from the city.[20] To the degree that this cleansing is accomplished, then to that degree there no longer exists a need for social welfare. Antihomeless laws are evidence of far more than "malign neglect" (Dear and Wolch 1996).

"Broken windows"

Advocates of 'public order' campaigns are forthright about this, and about their means for achieving purified space. Poor people are "disorderly" and they threaten the "quality of life" possible in the city (Kelling and Coles 1996; MacDonald 1995; Tier 1993; Wilson and Kelling 1982). Homeless people, 'quality of life' warriors assert, are like "broken windows" in a disinvested neighborhood (Wilson and Kelling 1982). They are an indication that a city space is "out of control." This "broken window" thesis is quite ubiquitous in American urban discourse. First given prominence in a 1982 *Atlantic Monthly* article by the criminologists James Q. Wilson and George L. Kelling, the broken window thesis argues that "arresting a single drunk or a single vagrant who has *harmed no identifiable person* seems unjust . . . [but] failing to do anything about a score of drunks or a hundred vagrants may destroy an entire community" (Wilson and Kelling 1982: 35, emphasis added). Wilson and Kelling's "broken window" argument suggests that even a single broken window in an urban neighborhood indicates a lack of care about urban space and invites other, more serious criminal behavior. A single broken window, Wilson and Kelling (1982: 31) argue, indicates that a building and surrounding property "become fair game for people out for fun and plunder."

Whatever the merits of these ideas as a theory of neighborhood decay (their roots, unsurprisingly, lay in the tenets about neighborhood control Jane Jacobs (1961) established), the broken window thesis is particularly dubious – and particularly offensive – when it is applied to people without homes to live in. "Untended *disorderly behavior*," like the broken window itself, Kelling (1987: 93, original emphasis) asserts, "also communicate[s] that nobody cares (or that nobody can or will do anything about disorder) and thus leads to an increasingly aggressive and criminal and dangerous predatory behavior." Wilson and Kelling (1982) are clear about what they mean by disorderly behavior: in addition to loitering teenagers, it is homeless people sitting on benches in parks, sleeping on heating grates, or passed out in alleys and curbsides. To Wilson and Kelling (1982: 29), such people are startlingly dehumanized (and thus their elimination from the space of the city is to be praised, not protested): "The citizen who fears the ill-smelling drunk, the rowdy teenager, or the importuning beggar," they write,

> is not merely expressing his distaste for unseemly behavior; he is also giving voice to a bit of folk wisdom that happens to be a correct generalization – namely, that serious crime flourishes in areas where disorderly behavior goes unchecked. *The unchecked panhandler is, in effect, the first broken window* ... If the neighborhood cannot keep a bothersome panhandler from annoying passers-by, the thief may reason, it is even less likely to call the police and identify a potential mugger or to interfere if the mugging takes place. (emphasis added)

Thus does the equation of people with broken windows become the justification for the legal and police actions we have already noted, not because homeless people are criminals, but because they are homeless.

The logic is incredible. Innocent people should be punished (or their actions should be made criminal), Wilson and Kelling are saying, because of the potential in a particular place for *other people's crimes* to occur. Or, more simply, homeless people should be criminalized because of who they are, rather than what they themselves have done.[21] At its boldest and its baldest, this defense of punitive measures against homeless people simply asserts that the *aesthetics of place* outweigh other considerations, such as the right of homeless people to find a means to live, to sleep, to *be* (Mitchell 1997b; Waldron 1991). As law scholar Steven Paisner (1994: 1272) argues in the midst of an attempt to develop constitutionally valid means of ridding city streets of homeless people, "the *most serious* of the attendant problems of homelessness is its devastating effect on a city's image" (emphasis added). Blowing up Pruitt-Igoe was not enough; now the people must be gotten rid of too. After all, nothing less than the city's "image" is at stake.

Indeed, George Kelling could not be clearer that his reason for advocating a broken window thesis is the elimination of people so as to preserve the aesthetics of place. "Quality of life" laws, he asserts, are not aimed at the homeless (he admits that homelessness is a "status"), but only at specific acts, specific forms of conduct. His argument is, quite simply, that "disorderly" behavior by homeless people (sleeping, urinating and defecating, loitering, begging) has nothing to do with their condition as homeless, but is rather behavior that is "voluntary" and can be controlled so as to enforce the "quality of life" in a city. He simply does not admit that they are *necessary* behaviors for those who have no other *space* in which to perform the everyday actions the rest of us take for granted (Waldron 1991). Writing with Catherine Coles 14 years after the *Atlantic Monthly* article, Kelling says that "the purpose of order maintenance is to prevent fear, crime, and urban decay ... order maintenance efforts are *not intended* to solve society's problems regarding" homeless people (by which they mean solving the problem of lack of housing: Kelling and Coles 1996: 222, emphasis added). Indeed, Kelling and Coles argue that order maintenance activities need to be framed as questions of *order* (conduct) and not homelessness (status). To even suggest that "quality of life" campaigns are about homelessness, Kelling and Coles (1996: 223) assert, leads to "an organizational, legal, and political trap" and should be avoided at all costs.

Kelling and Coles (1996: ch. 7) do agree that issues of housing and the homeless need to be addressed by cities. But precisely by decoupling homelessness from the question of public order (which focuses on the necessary actions of homeless people), by denying (against all evidence) that homeless people *must* perform necessary acts like sleeping, waste elimination, sitting, resting, and getting a living in public space, the "broken windows" thesis has been used to justify the most appalling of acts. When the city of Santa Ana, California, engaged in a quite brutal "deportation" of homeless people (to use the label the chief of police himself used) – a deportation that included holding homeless people against their will and with no charges filed against them in a municipal stadium and writing identification numbers on their bodies with indelible ink – the chief of police justified it in the pages of the *Los Angeles Times*' Orange County edition (Aug. 28, 1999) in an article called "Fixing Public's 'Broken Windows'" (see Simon 1992: 646 n. 96). In Seattle, city attorney Mark Sidrin relied on the "broken window" thesis to step up a campaign against homeless people in the early 1990s, at one point inviting George Kelling, who had gained prominence as the architect of both the New York Transit Authority's and the New York Police Department's campaigns against homeless people on trains, in stations, and on the sidewalks of the city (see Kelling and Coles 1996: 217), to meet with city council members, not in public session, but in private.

The point is that the "broken window" thesis is more than a thesis. Precisely by claiming that the maintenance of order is a separate issue from the problem of homelessness (despite the frequent use of it against homeless people by those who support it, including Kelling), it has become a primary justification for the removal of homeless people from urban public space. The logic works like this: first, the necessary behaviors of homeless people (sleeping, relieving themselves, sitting, loitering in public and, for some, begging) are outlawed in the name of "fixing broken windows." Second, since homeless people must therefore break the law when going about their everyday lives, they are, by definition, "disorderly" and "law breakers." Finally, city governments and much of the public come to the conclusion that the "problem lying on our sidewalks" is not a problem of homelessness, but a problem of public order, social norms, and aesthetics. In this manner, the "broken window" thesis becomes a primary tool in the war against homeless people, a tool, of course, that does nothing (because it is designed to do nothing) to address the underlying causes of homelessness.

"Broken windows" and urban diversity

Moreover, and rather incredibly, once the general logic of "broken windows" is accepted, the thesis can become an important tool for justifying the war against the homeless as a means of preserving *urban diversity*. Mark Sidrin, Seattle's city attorney, used the "broken window" thesis to argue that his goal was only to protect the rights of "the disabled, the elderly, the blind" and the homeless themselves who would become "victims of predators in their midst" if the sorts of behavior necessary to survival were not outlawed (quoted in Kelling and Coles 1996: 218). That is, he argued that "broken windows"-inspired anti-homeless laws serve as a chief means of preserving and promoting a vibrant and varied street life. The general council for the conservative American Alliance for Rights and Responsibilities, Robert Tier (1993: 286), is even more explicit on this point: "Current efforts to limit begging [and other actions of street people] are motivated by a desire to build and maintain a diverse, responsible, and interactive community, by maintaining and preserving vital public spaces where the community can interact" (emphasis added). Tier (1993: 287) puts the "broken window" thesis argument right at the center of his own reasoning and uses it to argue that new restrictions have to be placed on homeless people to thwart "the colonization of parks by people *wishing* to sleep and eat in the public place of one's own choosing, and to beg in any way one pleases" (emphasis added). True to Kelling and Cole's desire to separate such arguments from questions of the causes of home-lessness, Tier never once links the "wish" to sleep in a park to the *fact* that hundreds of thousands of people in America are forced to live without adequate shelter. And true to the new zeitgeist of the post-modern city in which universal rights are a thing of the past, Tier (1993: 286) promotes his vision of order in public spaces by arguing that those who fight for the rights of homeless people to have a place *to live* are "fighting yesterday's battles". In the punitive city, the postmodern city, the revanchist city (as Neil Smith (1996) calls it), diversity is no longer maintained by protecting, and struggling to expand, the rights of the most disadvantaged, but by pushing the disadvantaged out, making it clear that, as "broken windows" rather than people, they simply have *no* right to the city.[22] The language of diversity and difference is just as amenable to recidivist movements as it is progressive ones. And the solutions to social problems are hardly incompatible: if jail becomes the housing of first resort for homeless people now demonized as "broken windows," that should only please those urban theorists who, putting their faith in the liberatory power of 'heterotopia' see such spaces as places to be celebrated for the sorts of 'difference' they house. This is just what Smith (1998: 12) means when he argues that the disciplinary politics of neoliberalism is "welded into a proximate postmodern politics."

Yet it is still important to point out that this war against homeless people has not gone entirely unchallenged. Restrictions on begging have been challenged in court on the grounds that they violate the First Amendment (Mitchell 1998a); restrictions on sitting and loitering have been countered with arguments about the right to peaceably assemble; anticamping ordinances have been contested on the grounds that they constitute cruel and unusual punishment (because they punish status rather than conduct) and on the grounds that they interfere with the right to travel; and police sweeps of the homeless have been challenged on all these grounds but also because they constitute illegal searches and seizures (Mitchell 1998b).[23] Kelling's own campaigns to "take back the streets" and "take back the subways" of New York City have been legally challenged by advocates for the homeless on many of these grounds. In the New York cases, as nationally, the courts have been divided, with some upholding the rights of homeless people and others giving the police more-or-less *carte blanche* to rid the streets of home-less people.

Yet as legal scholar Robert Ellickson (1996: 1173) notes, the general trend in the courts has been to "encourage cities to formally zone their public space" so as to outlaw the activities of homeless people in some but not all places. The City of Miami, for example, has been ordered by a federal court to create "safe havens" for homeless people where they can be free from police harassment (*Pottinger* v. *City of Miami*, 810 F. Supp. 1551 [S.D. Fla. 1992]). In essence the plan calls for the ghettoization of homeless people, since it tacitly allows for police harassment in other parts of the city. Here is a new kind of 'heterotopia' in the making; here is the promotion of the "city of difference," post-justice version. Ellickson is worried about this process only to the degree that courts are "forcing" the zoning of public space on cities, rather than allowing them to develop through the desires and norms of localized communities themselves. He therefore suggests that the issues at stake in the war on homeless people are not actually ones of rights and social justice, so much as they are ones of what he calls proper "land management" (1996: 1171). It is worth looking at Ellickson's ideas in some detail because they represent a fascinating turn in the war on homeless people – a turn even more compatible with the vision of social justice promoted by theorists of the 'postmodern' city than even the "broken window" thesis.

The War on Homeless People, Round 2: Zoning Heterotopia

Depending on how you look at it, Robert Ellickson (1996), a stalwart of the conservative, property-based "law and economics" movement, presents a theory of urban heterotopia or a theory of the purification of urban space. As should be obvious in what follows (and from Foucault's own formulation), it is, in fact, invidious to distinguish between the two: both are an "impossible space" of social control with little room for the development of a rights-based notion of universalizing social justice. The Walter Meyer Professor of Property and Law at Yale University, Ellickson is perhaps best known for his book, *Order without Law: How Neighbors Settle Disputes* (1991), in which he makes the argument that "informal" controls on social order are often more effective than legal ones. In a nutshell, Ellickson is leery of the state, particularly the national state, but also to some degree more local jurisdictions too. He holds that the management of land needs to vary spatially in accordance with neighborhood, city, and state "norms," and that it should not be subject to universalizing federal constitutional oversight.

In 1996, relying explicitly on the "broken windows" thesis, Ellickson published what has become a quite influential law review article on what he calls "street disorder." Entitled "Controlling Chronic Misconduct in City Spaces: Of Panhandlers, Skid Rows, and Public Space Zoning,"[24] the article follows Kelling and Coles' (1996) injunction that questions of order in public space must be disconnected from issues of homelessness – by sheer force of will if no other way. Hence, though he contradicts himself in several places in the article (by pointing out, for example, that street people are in fact destitute and quite often homeless), Ellickson argues that "homelessness" is "an unduly ambiguous word" that "implies policy solutions that are inapt" (1192) and that "tends to entrap [homeless people] in a marginal status" (1193). Instead, he suggests that we should see "chronic misbehavior" on city streets as the product of two types of people: those he calls "bench squatters" (those who "monopolize" park or other benches and sidewalks with their bodies and belongings, be they homeless "bag ladies" or "Proust readers" (1184)); and those he calls "chronic pan-handlers" (those who beg in the same place day after day). Ellickson asserts that his policy proposals target these forms of *street conduct* and do not address the status of homelessness.

To the argument that antihomeless laws, because they target activities that homeless people have no choice in engaging in, target the involuntary *status* of being homeless rather than the specific acts they purport to regulate, Ellickson responds that "to treat the destitute as choiceless underestimates their capacities and, by failing to regard them as ordinary people, risks denying them full humanity" (1187). Begging, therefore, needs to be understood as "an option, not an inevitability" (1187). Indeed, according to Ellickson, beggars and benchsquatters are *more free* than the rest of us because, not "living lives structured around families and employers," street people have more time to "individually craft a daily routine" (1187) and "move from place to place" (1188). Ellickson even argues that ordering a mentally-ill woman "squatting" on a bench to "move along" "might actually enhance the liberties of the mentally ill" because "she herself might prefer that outcome to bearing the risks of involuntary confinement" (1189) in jail or an institution.[25] This all from a man who is already on record arguing that the provision of decent shelter to homeless people causes an increase in homelessness (Ellickson 1990).

Regulating public space: norms and harms

Ellickson's goal is to establish a program of public space regulation that does not rely on the universalizing tendencies of either law or rights, arguing that these "succumb to the notion that all open access spaces have to be governed by an identical regulatory regime" (1219 n. 301) and thus are inadequate to the differing needs of communities. They are spatially insensitive: "A constitutional doctrine that compels a monolithic law of public spaces," Ellickson intones, "is as silly as one that would compel a monolithic speed limit for all streets" (1247). Instead, he suggests that a spatially variable regime of urban public space zoning needs to be developed. A "city's codes of conduct," he argues, "should be allowed to vary spatially – from street to street, from park to park, from sidewalk to sidewalk" (1171–2). Optimally, this zoning should be "informal," that is, developed by the "community" as it establishes "norms" of behavior for people who use public spaces in its midst (1222–3). This informal zoning then should be maintained by "trustworthy police officers" (1173, 1245) that police to the norms the community has established. The second-best solution is for cities to create a formal system of public space zoning that allows for different sorts of behavior – and perhaps even some degree of "misconduct" – in the various public spaces of the city (1246).

Ellickson's target is "chronic street nuisance" which he defines as "behavior that i) violates community norms governing conduct in a particular public space ii) over a protracted period of time iii) to the minor annoyance of passers-by" (1175). That is, he is interested in regulating behavior that does no more than create a sense of "minor annoyance" – perhaps just a cracked, rather than a broken, window. Realizing that targeting behavior that is only of minor annoyance might not be generally palatable to policy-makers, Ellickson attempts to show that "chronic bench squatting" and non-aggressive "chronic panhandling" do in fact create a set of "harms" (1177) to the general community. First, these annoying behaviors may "trigger broken windows syndrome ... signal[ing] a lack of social control" (1177). Second, and more specifically, Ellickson notes that since many public authorities have taken to eliminating or redesigning public benches to discourage "bench squatting," a "proliferation of bench squatters ... tends to lead to the elimination of amply sized benches" from public space (1178 n. 50). (The direction of causality is not just insulting; it is symptomatic of a whole mode of reasoning concerning the homeless.) Third, Ellickson suggests that panhandling "worsen[s] race relations in cities where panhandlers are disproportionately black" (1181). (He also makes the offensive argument that one of the prime results of the Civil Rights movement was to make it easier for African Americans to live as homeless people on downtown streets.[26]) And fourth, according to Ellickson, "begging signal[s] an erosion of the work ethic," a "harm" that "all human societies" attempt to remedy (1182).

Given these "harms," Ellickson proposes the following rule: that "a person perpetrates a chronic street nuisance by persistently acting in a public space in a manner that violates prevailing community standards of behavior to the significant cumulative annoyance or persons of ordinary sensibility who use the same spaces" (1185). How should "community standards" be determined? Here Ellickson turns to Jane Jacobs: "The first thing to understand is that the public peace – the sidewalk and street peace," Ellickson quotes Jacobs (1961: 31–2) as saying, "is not kept primarily by the police, necessary as the police are. It is kept primarily by an intricate, almost unconscious, network of voluntary controls and standards among the people themselves and enforced by the people themselves" (1196). Jacobs' argument serves Ellickson well because he strives to show how what needs to be instituted are putative *community* norms. Yet since, as we will see, Ellickson has a remarkably truncated notion of who belongs in a community, he quickly discards Jacobs' argument in favor of promoting the police themselves as the primary guarantors of public order (1173, 1200–1, 1208–9, 1245). But his point in invoking Jacobs is not at all to debate the merits of city policing; rather, it is to deflect attention from that issue and to instead invoke a nostalgic vision of the city that serves as the template of the city of difference he wants to create.

Skid row: Ellickson's nostalgic heterotopia

This nostalgic vision is of a time when almost all American cities had within them what he calls "informally policed Skid Rows" (1208): the 1950s. In Ellickson's view Skid Row in the 1950s was a place "along with closely related Red Light Districts . . . where a city relaxed its ordinary standards of street civility" (1208). This is a quite partial view of Skid Row – the bulk of the evidence suggests it was always a heavily and stringently policed place (Anderson 1923; Bahr 1970, 1973; Bittner 1967; Blumberg et al. 1978; Foote 1956; McSheehy 1979; Wallace 1965; Wilson 1968) – but it does allow Ellickson to make a curious, if wholly unsupportable point that will become central to his whole argument: namely, that Skid Row not only was the appropriate home for alcoholics and the elderly poor, but made it possible for the police to act benevolently in their guaranteeing a *diversity* of social orders (1172, 1202–9). Here is what Ellickson says:

In Skid Row . . . moderate public drunkenness was likely to be tolerated, not only by the other down-and-out residents, but also by the police.[27] By contrast, the same level of inebriation elsewhere in downtown was much more likely to get an alcoholic in trouble. In the 1950s, a cop on the beat might unhesitatingly tell a "bum" panhandling or bench squatting in the central business district to "move along."[28] A bum on a Skid Row sidewalk would never hear this message because he was exactly where the cop wanted him.[29] In this way, the 1950s police officer helped to informally zone street disorder into particular districts. (1208–9)

Ellickson's own sources directly contradict him, showing how the police did in fact make frequent arrests on Skid Row, and tell the men and women there to "move along" (Foote 1956; Schneider 1986; Wilson 1968). And other sources, taking the ethnography of Skid Row into the 1970s (a period in which, Ellickson avers, policing of Skid Row was unduly hampered by constitutional restrictions on police power), show that the police could be impressively brutal in their use of arrest on Skid Row as a disciplinary mechanism (McSheehy 1979). But never mind.

For Ellickson, it is not the brutal policing of homeless men that was a problem; it could easily be excused on the grounds that it was less brutal than other parts of the city (1208–9 nn. 232–4). Rather, the "constitutional revolution" (1209) of the 1960s and 1970s that saw the overturning of law after law that criminalized status, Ellickson argues, made the sort of informal policing he champions impossible. Moreover, Ellickson complains, this "revolution" had the effect of "nationalizing" laws concerning street disorder (1209). That is, judicial liberalization and the recourse to constitutional law to litigate arrests for public drunkenness, vagrancy, loitering, and the like, applied a single standard of justice across all the urban spaces of the country. Such a "nationalization," in Ellickson's estimation, created a system that was "centralized and inflexible" (1213) making the sort of "informal zoning" that he thinks marked the 1950s Skid Row impossible (which, of course, was precisely the point). Thus, and also because churches suburbanized, Ellickson claims (1216), Skid Row fell into decline,[30] and the visible evidence of the "down-and-out" life diffused across the other spaces of downtown: "Street people who had previously been informally confined to Skid Row were now able to make chronic use of the busiest downtown areas. Many of them did" (1216).

Zoning public space: planning the postmodern heterotopia

Ellickson wants to return homeless and other poor people to Skid Row. Perhaps. In the most intriguing – but, as we will see below, thoroughly disingenuous – part of the article, a part that has attracted a good deal of attention from law scholars, policy analysts, and urban planners – Ellickson proposes that cities should be zoned into three categories – red, yellow, and green (1220–2). In Red Zones, which he argues should constitute perhaps 5 percent of a downtown area, "normal standards for street conduct would be significantly relaxed" (1221). "In these relatively rowdy areas," Ellickson (1221) writes, a city might decide to tolerate more noise, public drunkenness, soliciting by prostitutes, and so forth." Red zones would serve as "safe harbors for people prone to engage in disorderly conduct" (1221). Yellow Zones, which Ellickson thinks should cover 90 percent of downtown, should "serve as a lively mixing bowl" (1221). Here, "the flamboyant and the eccentric" (1221) would be allowed in just so long as they did not overstay their welcome. Here too *chronic* (but not episodic) panhandling and bench squatting . . . would be prohibited" (1221). Finally, Green Zones, occupying the final 5 percent of downtown space, would become "places of refuge for the unusually sensitive: the frail elderly, parents with toddlers, unaccompanied grade-school children, bench-sitters reading poetry" (but presumably not Proust) (1221–2). Even *episodic* panhandling and bench squatting would be outlawed.

In essence, then, Ellickson proposes to codify space such that at the scale of the city, a mix of "land uses" would be tolerated, and at the scale of the downtown 90 percent of the area would serve as a "lively mixing bowl" of peoples and activities, all overseen by a benevolent police force working to maintain the "community norms" of each area. His is precisely the vision of the postmodern, post-justice urban heterotopia. Within the discourse of the law, Ellickson is proposing a "third way" between strict antihomeless laws and the desires of those he calls "hyper-egalitarian" (1170), who presumably do not believe that urban space should ever be policed at all. Together, the red, yellow, and green zones represent a concretization of Soja's (1996) "Thirdspace" – that space that exists somewhere between Lefebvrian "first space" of spatial practices (the homeless doing what they must) and the "second space" of orderly representation and control (the aims of the antihomeless laws). Ellickson provides a means to *achieve* the "space of extraordinary openness" that Soja advocates. That his motive might be different from Soja's is, quite literally, immaterial.

Echoing in the realm of constitutional law arguments frequently made in postmodern urban theory, Ellickson advocates the need for openness and flexibility, arguing that "federal constitutional rulings are one of the most centralized and inflexible forms of lawmaking. In a diverse and dynamic nation committed to the separation of powers and federalism, there is much to be said for giving state and local legislative bodies substantial leeway to tailor street codes to city conditions" (1213–14). Ellickson approvingly quotes Supreme Court Justice Hugo Black (*Powell* v. *Texas*, 392 US 514, 547–8):

It is always time to say that this Nation is too large, too complex and composed of too great a diversity of peoples for any one of us to have the wisdom to establish the rules by which local Americans must govern their local affairs. (1248)

Reflecting the postmodern concern that universal standards, "no matter how progressive," as Soja (1997) puts it, are oppressive of "difference," advocates of the ghettoization of homeless people declare universalizing standards of justice – concerning the right to be on and to use streets, for example – to ignore the differing needs of the local. As in postmodern calls for spatial politics based on the disordering power of difference, Ellickson's "informal zoning" of public space allows cities to "spatially differentiate their street policies" (1219).

Zoning for whom?

But such a position simply begs the key questions, the answers to which are deeply troubling both in Ellickson's, and in postmodern theory's, vision of heterotopia; namely: *Which* local people shall establish these spatially sensitive policies and practices; and *How – under what conditions and procedures* – should they do so?

For his part, Ellickson argues that the zoning of public space should be done informally by "members of a close-knit group who repeatedly make use of open access public space" (1222). These members should "enforc[e] social norms to deter an entrant from using [public space] in a way that would unduly interfere with the opportunities of other members" (1222). Such community members, according to Ellickson, are particularly adept at "recogniz[ing] the crazy-quilt physical character of urban spaces and the myriad demands of pedestrians [and they] tend to vary their informal norms from public space to public space" (1222). In this effort, they are aided by the police who work to enforce these varying community norms (1223). And over time, Ellickson hopes, residents and homeless people alike will internalize these rules, and the norms of the different zones will become second nature (1225–6). Yet, even so, Ellickson recognizes that informal zoning might not be effective (and here, perhaps, he parts company with those who advocate a more radical form of postmodern planning). Though he doesn't say it outright, it is clear that Ellickson is concerned that "chronic panhandlers" and "bench squatters" could be taken for *members* of the "close-knit group" that uses public space, and so he suggests that city governments should be given leeway to formally zone public space into red, yellow, and green zones, replete with signs listing applicable rules.

Despite this worry about the efficacy of informal zoning, Ellickson still puts his faith in "the community" for whom public space will be policed. Yet, and this is crucial, Ellickson *never* explicitly defines community. What emerges, in the course of his long law-review article, however, is that this "community" simply does not include homeless people. They are in no sense considered to have any rightful claim to the use of streets and parks – they are figured only as unwanted nuisances. They have no standing whatsoever as members of the communities in which they live. So who then is included in this community? In the only hint at the community he has in mind, Ellickson points to various "individual champions of the public": pedestrians, owners and occupiers of abutting land, and organizations that enforce street decorum (like Business Improvement Districts and the police) (1196–9). With the exception of "orderly" pedestrians (1197) and the police who work in the interest of the "community" as a whole, Ellickson's community is thus apparently the community of *property*, since, as he shows, it is property that suffers the greatest "harm" from nearby homeless people engaged in little more than "minor annoyances."

Yet Ellickson must recognize (since much of his research has been on issues of zoning and land management) that this community of property (including renters) will fight against red-zone designation: no community would willingly accept such a status as it would incur unacceptable costs in terms of falling property values and increased maintenance and service costs. The "negative externalities" attendant upon the creation of an official ghetto (with its designating signs) would be too great. His advocacy of "informal" zoning, therefore, stands as all the more curious and fanciful: he never addresses the question of why informal zoning would not suffer the same problem as formal zoning; why would property owners (or adjacent property owners at the edges of the district) not resist the decline of property values attendant upon the harboring of "broken windows." The answer is they would not,[31] and thus the development of freely tolerated – not judicially mandated – "red zones" is extremely unlikely, especially since, in Ellickson's view "the first best solution to the problem of street misconduct would be the maintenance of a trustworthy police department, whose officers would be given significant discretion in enforcing *general standards* against disorderly conduct and public nuisances" (1245, emphasis added). Ellickson never even broaches the question of how "general standards" and "community norms" are determined. In Ellickson's proposal there simply is no mechanism – and certainly no *democratic* mechanism – outlined for instituting informal zoning, much less guaranteeing any level of spatial justice. Not coincidentally, then,

Ellickson suggests absolutely no means – legal, constitutional, legislative – for guaranteeing that *any* space would be zoned red.

And this is precisely *why* Ellickson so hopes to win approval for his plan for informal zoning. It is, in the end, an elaborate hoax, but an extremely dangerous one: it leads to exactly the same hoped-for outcome as antihomeless laws in general: the elimination of not homelessness but homeless people, by eliminating any space in which they can *be*. The differentiation of space, this practical plan for implementing a "lively mixing bowl" in the city, *is* nothing more than the desire for the same purified space that sits behind antihomeless laws, but with this crucial difference: to the degree that Ellickson's plans for informal zoning are adopted, they will remain out of the purview of the courts, making it that much easier to eradicate homeless people. And it is for just that reason that I have – perhaps invidiously – aligned Ellickson's proposals with key arguments about planning "in and for postmodernity" and about the nature of postmodern urbanism: by their very distrust of universalizing justice, such arguments eliminate one of the most important weapons that those who seek a *progressive* city have in this war on homeless people: the appeal to right, the appeal to rights, the appeal to what is universally *just*.

Conclusion: The Postmodern Impulse and the Post-Justice City

> Postmodernism lives. Legions of detractors and years of intellectual debate have done nothing to arrest its expansion or reduce its impact. . . . Despite or because of being profanely ambivalent and ambiguous, rejoicing in consumption or celebrating obsessions, ignoring consistency and avoiding stability, favoring illusion and pleasure, postmodernism is the only possible answer to a century worn out by the rise and fall of modern ideologies, the pervasion of capitalism, and the unprecedented sense of personal responsibility and individual impotence.
>
> Whether one likes it or not, postmodernism is a state of things. It is primarily determined by an extremely rapid and freewheeling exchange to which most responses are faltering, impulsive, and contradictory. What is at stake is the very constitution of being – the ways we perceive ourselves and others, the modes of experience that are available to us, the women and men whose sensibilities are shaped by urban exposure.
>
> *Celeste Olalquiaga 1992: xi, quoted in Soja 1996: 92*

It will take imagination and political guts, a surge of revolutionary fervour and revolutionary change (in thinking as well as in politics) to construct a requisite poetics of understanding for our urbanizing world, a charter for civilization, a trajectory for our species being, out of the raw materials of the present.

David Harvey 1996: 438

When "revealed from below," that is, when examined from the perspective of its effects on the poorest and most marginalized residents, the "postmodern city" proves not to be the outcome of some "apparently random" set of processes, but the precise outcome of concerted, painstakingly made plans: the dismantling of welfare; the retreat from social housing; the reconfiguration of policing so as to enforce marginalization; the promotion of plans to annihilate those residents deemed to be the human equivalent of blight (see Fox Piven 1998; Mair 1986). Just as troubling, however, is the degree to which postmodern urban theory, at least as it is represented in the work of prominent geographers like Soja, Dear, and Flusty, is fully inadequate to addressing the *needs* of such residents. After all, as postmodernists never tire of pointing out, the postmodern world is no longer needs-focused; that was a hallmark of modernism. Now it is *desire* that matters. Yet it should be deeply troubling, at the very least, that the desires of the well-to-do – that community of property that Ellickson speaks for – are so carefully planned for (by refocusing police practices, by promoting certain kinds of zoning); not despite the ongoing needs of homeless and other poor people, but *instead of them.*

For that is what is at stake in the "post-justice" city, the city no longer defined by the struggle for social justice: no longer is decent housing on the agenda (as one front for progressive change among many), but now it is a question of the best way to exterminate homeless people. While antihomeless laws seek the annihilation of homeless people through the expedient of removing any space where they can possibly *be,* any space where they can perform the functions necessary to sustaining life (Mitchell 1997b; Waldron 1991), Ellickson's advocation of a more subtle informal zoning does exactly the same thing by projecting *private property interests* into public space. Ellickson's goal, like that of the architects of "quality of life" campaigns, is nothing less than the total elimination of homeless people from the spaces of the city.

The only thing that makes it possible for homeless people to live is access to public space governed by no private property law. Anti-homeless laws and the zoning of public space alike seek to eliminate just those spaces. The ugly reality of the postmodern American city, the city after Pruitt-Igoe that no longer even pretends to provide decent housing for its residents (nor even enough temporary shelter space), is that streets and sidewalks, empty lots and parking lots, parks and squares, are the

only home for hundreds of thousands of people. It is also the ugly reality
of the postmodern city that *progressive* homeless politics has been
reduced to arguing for the right to beg, the right to shit in a park, or
the right to sleep under a bush. The problem with rights-based arguments for universal standards of social justice – those standards partially made concrete in public housing – is not that they were universal,
but that they were not universal *enough*. The dream of decent affordable housing turned into something of a nightmare because it was never
completed, because it did not expand far and deep enough into society.
It is time to revive the struggle for universal and universalizing rights to
housing in the city. Heterotopia is simply no substitute for a house and
home.

To revive that struggle, however, a key weapon in the postmodern
urban arsenal must be neutralized (a job for which geographers are particularly well-suited), namely, *spatialization*. For as Berger (1974) and
Soja (1989) rightly remark, and as Ellickson intuits, it is *space* that hides
consequences from us. Ellickson's proposals are particularly pernicious
because, in congruence with proponents of postmodern urbanism, by
presenting the postmodern city as the city of (disordered) difference
rather than a city where universal rights are to be struggled over and
won, by *spatializing* homelessness, rather than understanding it as the
social and *economic* problem that it is, it, like so much postmodern
theory, deflects attention from what must be done to eliminate homelessness, including constructing safe, democratic *public* housing and
revitalizing a *public* and universal commitment (and public financing
for) sensitive mental health intervention.[32] To abandon that commitment, to assume along with Spivak that all struggles are simply local
(because capitalism is so efficient at localizing effects), or that, along
with Dear and Flusty, that the urban mosaic is like a "keno card," to
argue for a form of politics that celebrates marginality rather than
seeking to redress it – that is, in each case to follow Ellickson and substitute the spatial for the social – is folly. The postmodern impulse has
little constructive, little progressive, to say about the post-justice city:
to my mind, by its refusal to recognize the *necessity* of universal and
universalizing rights and modes of knowing and being, it is quite com-
plicit with it.

Understanding the spatiality of the contemporary city is certainly
of pressing importance. But unless that spatiality is linked directly to
a clear vision of social justice that takes basic human rights, such as
to shelter and livelihood, as universal and non-negotiable, even if that
means undermining some of the 'difference' postmodern urban theorists
so highly value, it will continue to be nothing more than "profanely
ambivalent and ambiguous, rejoicing in consumption [and] celebrating

obsession." Contemporary social theory, and more importantly, contemporary politics, requires a careful attention to 'postmodern geographical praxis' not as it is worked out in the books and journals of
academics, but as it is worked out on the streets themselves, in the government chambers and policy thinktanks that are the front lines of the
war on the homeless, and in the courthouses that have often proven
amenable to undermining the basic rights and needs of so many people
in order to protect the desires of the privileged.

Linking this analysis of the spatial more closely to the social becomes
doubly important when we realize how easily Jane Jacobs' (1961) attack
on modernist planning has been transmogrified into a brutal urban
regime where it becomes quite unremarkable to argue that poor people
are only so many 'broken windows,' and that the solution to these frac-
tures is not to repair them but to remove them altogether. This is
the Pruitt-Igoe solution: don't repair and reinvest, eliminate. This is the
post-justice solution. This is the solution that *in fact* determines the
shape and structure of the postmodern city: it is what is behind Dear
and Flusty's "apparent" randomness. Under this regime, the American
city has become a place where regulation is supple and responsive to
the needs of the rich, the middle classes, and the desires of property.
It is the city where pleasure and fun (for some) always trump material needs (for all) and we are all supposed to celebrate that fact (as
Olalquiaga makes clear). In this sense Wolch and Dear's analysis in
Malign Neglect (1996) is right on target.

If we want to know what postmodern geographical praxis is – if we
want to resist the postmodern impulse in the academy that is so com-
plicit with it – then we would do well to look carefully not at the urban
landscape as if it is a keno card or a species of "hyperreality" that simply
cannot be unmasked (Soja 1997: 244), but at those actors and those
policies that are actively structuring it: the broken windows policies of
a George Kelling, or the antihomeless zoning of a Robert Ellickson, and
to trace how they have been implemented by people like Seattle's Mark
Sidrin and New York's Rudolph Giuliani.

Olalquiaga's depressing diagnosis is in many senses correct: postmodernism is a state of things (brought into being by, among so many
other things, the sort of praxis outlined in this chapter), but it is one
that, along with the genocidal politics of antihomelessness ought to be
struggled against, especially by those of us who profess any concern
whatsoever for those who truly are "shaped by urban *exposure*." It's
time to resist this complacent postmodern impulse and rely instead
on "imagination and political guts." Our goal should be precisely
a search for order and stability – *progressive* order, and equitable
stability: a clear, straightforward vision of universal social justice

must be reconstructed in these "post-justice" times. The hundreds of thousands of people sleeping on the streets every night deserve nothing less.

Acknowledgments

This chapter has been presented, in a variety of forms, to the Graduate Students' Conference, Ohio State University, October 1997; Queen's University Department of Geography, February 1998; the University at Buffalo, February 1998; the Cornell Coalition on the Homeless, April 1998; the Syracuse Socialist Forum, November 1998; and the Postmodern Geographical Praxis Conference, Venice, June 1999. I thank the organizers and audience at each of these venues for the opportunity and for probing questions. Special thanks to Neil Smith and Cindi Katz for encouragement and criticism; and to Claudio Minca for prodding me to assure that my aim was on the proper target. He bears no responsibility for any of my shots that are nonetheless errant.

NOTES

1   I take these oppositions, which can be found ubiquitously in the literature commenting on the "postmodern turn," from table 10.1 in Cuthbert (1995) and from table 1.1 in Harvey (1989a).
2   Dear and Flusty (1998: 62) admit that the "apparently random" nature of urban development and redevelopment, is "determined by a rationalized set of procedures beyond the territory of the card itself," but they never examine in any detail what these processes might be. More problematic, however, is that such a theoretical position, one that takes randomness as *a priori* the determining factor of landscape change, turns attention from the actual socio-spatial practices, operating at several scales (including the scale of the "card" itself) that can, as we will see, be empirically identified. Any "apparent" randomness is just that: apparent. *Assuming* randomness, however, merely deflects attention from the practices – and the structures of culpability – that produce the landscape.
3   "Praxis" is generally taken to mean critically-informed practice; more expansively, it refers to the unleashing of human creativity. A keyword in the Marxist lexicon, in this essay I am going to give it a slightly different twist. It seems to me that those who actually plan the oppressive and regressive city that has arisen in the postmodern era are the ones most effectively engaged in postmodern praxis: their practices are nothing if not vitally critical, their plans viciously creative.
4   Perhaps a hallmark of this new geographical praxis is a disinterest in logical consistency. At the outset of their paper Dear and Flusty (1998:

50–1) point out that discussions as to whether a "break" has occurred are "enervating" and that it is more useful just to *assume* that one has and to precede from there. Fifteen pages later, however, that *assumption* has become an established *fact*: "*our investigation has uncovered an epistemological radical break* with past practices" (65, original emphasis). Nowhere is it clear how a starting assumption became a point proven, nor is it ever clear why such a move is itself not a contribution to the enervation of debate.

5   Harvey uses this term to indicate the way that social structures, institutions, built forms, and so forth, are always a product of ongoing struggle, contestation, cooperation – in short, social relations, taking his cue from the dialectical reasoning of Marx (particularly as phrased by Ollman 1993), as well as the "new sociology of science" of, for example, Latour (1987) and Callon (1986), and the feminist reinterpretations of knowledge and practice of Haraway (1989, 1991).

6   Soja's work itself is something of a "Thirdspace," having moved from a position on postmodernism not too dissimilar to Harvey's (cf. Soja 1989) to one now much closer, and perhaps far more radical, than Dear and Flusty's.

7   As Neil Smith (1998: 17) puts it, "The homily that 'all politics is local' is fatuous and self-defeating in the face of global neo-liberalism. Not that the local is irrelevant, far from it; only that it is not exclusionary: 'act and think locally *and* globally' is a far better guide to action."

8   My point is not at all to defend the design of Pruitt-Igoe, nor to deny that there was much about the project and US public housing policy as a whole that was incredibly dehumanizing. Rather it is to point to the fact that the destruction of Pruitt-Igoe symbolized not a renewed commitment to social justice, but its eclipse.

9   For an accessible history that effectively places housing policy within a larger theory of social movements and the capitalist state see Piven and Cloward (1993).

10   Though they do, certainly, appear in Dear and Flusty's (1998) accounting of postmodern urban form, even if a decent explanation of their persistence in the postmodern era does not.

11   The title is taken from Ruddick's (1990) provocative article; see also Ruddick (1996: ch. 3).

12   The capitalist state is, of course, contradictory. While it is often a force of oppression serving the interests of capital and privilege, it is also the primary guarantor of human and social rights (see Mitchell 1997a). This latter function is the one that needs to be expanded, *at the expense of* the former; yet it is this latter that has been abandoned by many contemporary "critical" social theorists.

13   The term "compassion fatigue" was coined in the late 1980s to describe the apparent public withdrawal of support for homeless people. The argument is that the seeming intractability of homelessness – and the seeming permanence of street people asking for handouts – had worn out even the most liberal members of the public.

14   As Daniel Singer (1999) argues, much of the contemporary left has simply accepted the Thatcherite dictum that "There is No Alternative" to capitalist domination and the discipline of the market. This has had profound effects on the nature of social struggle in the city (as elsewhere).

15   Genocchio (1985: 42) suggests that "simply to write off his conception of heterotopias as incoherent seems less than prudent." Yet his own argument in favor of the term, an insupportably idealist notion that heterotopia "is an idea which . . . produces space as transient, contestatory, plagued by lapses and ruptured sites," suggests that to retain it is to retreat from the realm of social practice and to place faith in "theory" as the producer of a just world. That hardly makes for a strong argument for retaining an idea despite its admitted "incoherence."

16   The numbers reported here are from the US Department of Housing and Urban Development and the US Bureau of the Census, as reported on the National Law Center on Homelessness and Poverty homepage (www.nlchp.org/h&pusa.htm), accessed August 17, 1999. These numbers, by many indications, are now outdated and underrepresent the extent of homelessness. Despite a record low official unemployment rate in the last years of the 1990s, homelessness continues to grow. Shelter operators report that they are seeing more people, and people are staying longer in shelters, than even five years ago. In New York state, shelter operators attribute this to the "end of welfare" (News report, WRVO Radio (Oswego), August 18, 1999).

17   I make no effort in this paper to explain the inexplicable disconnection of the careful empirical and socially-infused research of Michael Dear on homelessness, which exhibits all the hallmarks of a deep commitment to ontological and epistemological determinacy, and his more philosophical and speculative treatises on postmodern urban space and planning, which seek to promote ontological and epistemological indeterminacy.

18   Outlawing urinating and defecating in public is eminently reasonable until one remembers that American cities, as a rule, do not provide publicly accessible toilets. Likewise, enforcing a ban on sleeping in public may seem reasonable, except to those for whom there is simply nowhere else to sleep, as is the case for the *at least* 425,000 people each night who are homeless and exceed the available shelter space (figures from ICH 1994: 40; see Foscarinis 1996: 13).

19   Kelling and Coles (1996), in their influential book on public order, explicitly make this argument.

20   I use the term "cleanse" purposely, first because of the nearly ubiquitous association of homelessness with dirt and decay; and secondly to hint at the parallel efforts to create purified spaces under the rubric of "ethnic cleansing." The desire for purified space is explored in Cresswell (1996) and Sibley (1995).

21   Beginning in the late 1950s, shifts in American constitutional law led to the repeal of most "status crime" laws: those, like vagrancy laws, that punished people for an involuntary status rather than some identifiable

conduct. In essence, Wilson and Kelling are calling for the return of status crime punishment. For a geographical analysis see Mitchell (1998b).

22 How different this is to the arguments made by Lefebvre (1996) in support of his rallying cry: the "right to the city." Perhaps at this point, in the context of the arguments about diversity, I should note that the most punitive measures against the homeless coincide with an increasing "African-Americanization" of the homeless population, a point that will be picked up in the next section.

23 Kelling, once an advocate of police sweeps of the homeless, now argues against them on the grounds that they create too much adverse publicity (Kelling and Coles 1996: 224).

24 All unaccompanied page numbers in this section refer to this article.

25 The nature of this "freedom" was made plain in November 1999, when New York's Mayor Giuliani ordered city police to arrest and jail any homeless person in city streets or parks who refused to move along when ordered to do so. As Sartre once commented, he was never so free than that moment on a Paris street when a Nazi soldier held a gun to his head and told him to cross the street.

26 "The softening of white hostility towards blacks during and after the 1960s seems to have allayed the reservation many underclass blacks had previously harbored about becoming chronic users of downtown spaces. In any event, the panhandlers and street homeless who began appearing in American downtowns after 1980 were disproportionately black. No fact better demonstrates the success of the post-1960 inclusionary zeitgeist" (1216–17).

27 Ellickson notes in a footnote at this point that police *did* in fact frequently arrest drunks on Skid Row – and that they engaged in regular "sweeps" of the streets (i.e. they indiscriminately detained or arrested street people on Skid Row), but dismisses this evidence with the comment that even so the police seemed to be "more permissive" on Skid Row than in other parts of town (1208 n. 232, citing Bittner 1967). The evidence does not support Ellickson (see, for example, McSheehy 1979).

28 Here Ellickson adds a footnote saying, "One can only conjecture how often night-sticks were used to enforce these orders." Actually, one could read the ethnographic and historical evidence, including Ellickson's own sources (e.g. Bahr 1973; Bittner 1967; Wilson 1968) (1208 n. 233).

29 And here Ellickson's own footnote directly contradicts the message in the body of the paper. Where in the body Ellickson says "a bum would never hear this message," in the footnote he points out just how frequently they did, but once again says this does not matter since police were "more tolerant" in Skid Row than elsewhere. Thus some rather brutal policing tactics are justified because they are not as brutal as they could conceivably be (1209 n. 234).

30 Ellickson does not examine processes of gentrification, assumes that urban renewal followed Skid Row decline, and is skeptical of the role of SRO destruction in the growth of homelessness (1216 n. 279).

31   The best analogy is probably the history of both formal and informal zoning against adult cinemas, bookstores, and strip-clubs, up to Giuliani's recent campaign to rid Manhattan of all "adult" establishments, despite formal zoning laws allowing them.
32   The State of New York has now moved in the opposite direction: in August 1999, the Governor signed a law criminalizing those with mental illness who fail to take the prescribed medication.

REFERENCES

Anderson, N. (1923). *The Hobo: The Sociology of the Homeless Man*. Chicago: University of Chicago Press.

Bahr, H. (1970). *Disaffiliated Man: Essays and Bibliography on Skid Row, Vagrancy, and Outsiders*. Toronto: University of Toronto Press.

Bahr, H. (1973). *Skid Row: An Introduction to Disaffiliation*. New York: Oxford University Press.

Berger, J. (1974). *The Look of Things*. New York: Viking.

Bittner, E. (1967). The Police on Skid Row: A Study of Peace Keeping. *American Sociological Review* 32: 699–715.

Blumberg, L., T. Shipley, and S. Barsky (1978). *Liquor and Poverty: Skid Row as a Human Condition*. New Brunswick: Rutgers Center of Alcohol Studies.

Callon, M. (1986). Some Elements of a Sociology of Translation: Domestication of the Scallops and Fisherman of St. Brieuc Bay. In *Power, Action and Belief*, ed. J. Law. London: Routledge and Keegan Paul.

Cox, K. and A. Mair (1989). Levels of Abstraction in Locality Studies. *Antipode* 21: 121–32.

Cresswell, T. (1996). *In Place/Out of Place: Geography, Ideology and Transgression*. Minneapolis: University of Minnesota Press.

Cuthbert, A. (1995). Under the Volcano: Postmodern Space in Hong Kong. In *Postmodern Cities and Spaces*, eds. S. Watson and K. Graham. Oxford: Blackwell, 138–48.

Davis, M. (1991). *City of Quartz: Excavating the Future in Los Angeles*. London: Verso.

Dear, M. and S. Flusty (1998). Postmodern Urbanism. *Annals of the Association of American Geographers* 88(1):50–72.

Dear, M. and J. Wolch (1987). *Landscapes of Despair*. Princeton: Princeton University Press.

de Certeau, M. (1984). *The Practice of Everyday Life*. Berkeley: University of California Press.

Ellickson, R. (1990). The Homelessness Muddle. *Public Interest* 99: 45–60.

Ellickson, R. (1991). *Order Without Law: How Neighbors Settle Disputes*. Cambridge: Harvard University Press.

Ellickson, R. (1996). Controlling Chronic Misconduct in City Spaces: Of Panhandlers, Skid Rows, and Public Space Zoning. *Yale Law Journal* 105: 1165–248.

Foote, C. (1956). Vagrancy-Type Law and Its Administration. *University of Pennsylvania Law Review* 104: 603–50.

Foscarinis, M. (1996). Downward Spiral: Homelessness and Its Criminalization. *Yale Law and Policy Review* 14: 1–63.

Foucault, M. (1970). *The Order of Things: An Archaeology of the Human Sciences*. London: Tavistock.

Foucault, M. (1986). Of Other Spaces. *Diacritics* 16: 22–7.

Genocchio, B. (1995). Discourse, Discontinuity, Difference: The Question of 'Other' Spaces. In *Postmodern Cities and Spaces*, eds. S. Watson and K. Graham. Oxford: Blackwell, 35–46.

Gibson-Graham, J.-K. (1996). *The End of Capitalism (As We Knew It)*. Oxford: Blackwell.

Haraway, D. (1989). *Primate Visions: Gender, Race and Nature in the World of Modern Science*. New York: Routledge.

Haraway, D. (1991). *Symians, Cyborgs, and Women: The Reinvention of Nature*. New York: Routledge.

Harvey, D. (1989a). *The Condition of Postmodernity: An Enquiry into the Origins of Cultural Change*. Oxford: Blackwell.

Harvey, D. (1989b). *The Urban Experience*. Baltimore: Johns Hopkins University Press.

Harvey, D. (1996). *Justice, Nature and the Geography of Difference*. Oxford: Blackwell.

ICH (Interagency Council on the Homeless) (1994). *Priority: Home! The Federal Plan to Break the Cycle of Homelessness*. Washington, DC: Department of Housing and Urban Development.

Jacobs, J. (1961). *The Life and Death of Great American Cities*. New York: Random House.

Jencks, C. (1981). *The Language of Postmodern Architecture*. New York: Rizzoli, 3rd edn.

Katz, C. (1998a). Excavating the Hidden City of Social Reproduction: A Commentary. *City and Society*. Annual Review, 37–46.

Katz, C. (1998b). Whose Nature, Whose Culture? Private Productions of Space and the 'Preservation of Nature.' In *Remaking Reality: Nature at the Millennium*, eds. B. Braun and N. Castree. New York: Routledge, 46–63.

Kelling, G. (1987). Acquiring a Taste for Order: The Community and Police. *Crime and Delinquency* 33: 90–102.

Kelling, G. and C. Coles (1996). *Fixing Broken Windows: Restoring Order and Reducing Crime in Our Communities*. New York: The Free Press.

Kodras, J. (1997). Restructuring the State: Devolution, Privatization, and the Geographic Redistribution of Power and Capacity in Governance. In *State Devolution in American: Implications for a Diverse Society*, eds. L. Staeheli, J. Kodras, and C. Flint. Thousand Oaks: Sage Publications, 79–96.

Kunstler, J. (1993). *The Geography of Nowhere: The Rise and Decline of America's Man-Made Landscape*. New York: Simon and Schuster.

Lash, J. and D. Urry (1987). *The End of Organized Capitalism*. Oxford: Blackwell.

Latour, B. (1987). *Science in Action: How to Follow Scientists and Engineers through Society*. Cambridge: Harvard University Press.

Lefebvre, H. (1996). *Writings on Cities*, selected, transl., and introduced by E. Kofman and E. Lebas. Oxford: Blackwell.

Ley, D. (1996). *The New Middle Class and the Remaking of the Central City*. Oxford: Oxford University Press.

MacDonald, H. (1995). San Francisco's *Matrix* Program for the Homeless. *Criminal Justice Ethics* 14(2):79–80.

Mair, A. (1986). The Homeless and the Post-Industrial City. *Political Geography Quartery* 5: 351–68.

McCann, E. (1994). Neotraditional Developments: The Anatomy of a New Urban Form. *Urban Geography* 13: 210–33.

McSheehy, W. (1979). *Skid Row*. Boston: G. K. Hall and Cambridge: Schenkman Publishing.

Mitchell, D. (1997a). State Restructuring and the Importance of 'Rights Talk.' In *State Devolution in American: Implications for a Diverse Society*, eds. L. Staeheli, J. Kodras, and C. Flint. Thousand Oaks: Sage Publications, 7–38.

Mitchell, D. (1997b). The Annihilation of Space by Law: The Roots and Implications of Anti-Homeless Laws in the United States. *Antipode* 29: 303–35.

Mitchell, D. (1998a). Anti-Homeless Laws and Public Space I: Begging and the First Amendment. *Urban Geography* 19: 6–11.

Mitchell, D. (1998b). Anti-Homeless Laws and Public Space II: Further Constitutional Issues. *Urban Geography* 19: 98–104.

NLCHP (National Law Center on Homelessness and Poverty) (1991). *Go Directly to Jail: A Report Analyzing Local Anti-Homeless Ordinances*. Washington: National Law Center on Homelessness and Poverty.

NLCHP (National Law Center on Homelessness and Poverty) (1993). *The Right to Remain Nowhere: A Report on Anti-Homeless Laws and Litigation in 16 United States Cities*. Washington: National Law Center on Homelessness and Poverty.

NLCHP (National Law Center on Homelessness and Poverty) (1994). *No Homeless People Allowed: A Report on Anti-Homeless laws, Litigation and Alternatives in 49 United States Cities*. Washington: National Law Center on Homelessness and Poverty.

NLCHP (National Law Center on Homelessness and Poverty) (1996). *Mean Sweeps: A Report on Anti-Homeless Laws, Litigation and Alternatives in 50 United States Cities*. Washington: National Law Center on Homelessness and Poverty.

Olalquiaga, C. (1992). *Megalopolis*. Minneapolis: University of Minnesota Press.

Ollman, B. (1993). *Dialectical Investigations*. New York: Routledge.

Paisner, S. (1994). Compassion, Politics, and the Problems Lying on Our Sidewalks: A Legislative Approach for Cities to Address Homelessness. *Temple Law Review* 67: 1259–305.

Piven, F. F. (1998). Organizing the Poor for the Year 2000. Paper Presented to the Syracuse Social Movements Initiative, Syracuse University, Oct. 29.

Piven, F. F. and R. Cloward (1977). *Poor Peoples' Movements: Why They Succeed, How They Fail.* New York: Pantheon.

Piven, F. F. and R. Cloward (1993). *Regulating the Poor: The Functions of Public Welfare,* updated edn. New York: Vintage.

Ruddick, S. (1990). Heterotopias of the Homeless: Strategies and Tactics of Place Making in Los Angeles. *Strategies: A Journal of Theory, Culture and Politics* 3: 184–201.

Ruddick, S. (1996). *Young and Homeless in Hollywood: Mapping Social Identities.* New York: Routledge.

Schneider, J. (1986). Skid Row as an Urban Neighborhood, 1880–1960. In *Housing the Homeless,* eds. J. Erickson and C. Wilhelm. New Brunswick: Center for Urban Policy Research, 67–189.

Sibley, D. (1995). *Geographies of Exclusion.* London: Routledge.

Simon, H. (1992). Towns Without Pity: A Constitutional and Historical Analysis of Official Efforts to Drive Homeless People from American Cities. *Tulane Law Review* 66: 631–76.

Simon, H. (1995). The Criminalization of Homelessness in Santa Ana, California: A Case Study. *Clearinghouse Review* 29: 725–9.

Singer, D. (1999). *Whose Millennium? Theirs or Ours?* New York: Monthly Review Press.

Smith, N. (1996). *The New Urban Frontier: Gentrification and the Revanchist City.* New York: Routledge.

Smith, N. (1998). Giuliani Time: The Revanchist 1990s. *Social Text* 57: 1–20.

Soja, E. (1989). *Postmodern Geographies.* London: Verso.

Soja, E. (1996). *Thirdspace: Journeys to Los Angeles and Other Real-and-Imagined Places.* Oxford: Blackwell.

Soja, E. (1997). Planning In/For Postmodernity. In *Space and Social Theory: Interpreting Modernity and Postmodernity,* eds. G. Benko and U. Strohmeyer. Oxford: Blackwell.

Sorkin, M., ed. (1992). *Variations on a Theme Park: The New American City and the End of Public Space.* New York: Hill and Wang.

Tier, R. (1993). Maintaining Safety and Civility in Public Spaces: A Constitutional Approach to Aggressive Begging. *Louisiana Law Review* 54: 285–338.

Till, K. (1994). Neotraditional Towns and Urban Villages: The Cultural Production of a Geography of 'Otherness.' *Environment and Planning D: Society and Space* 11: 709–32.

Waldron, J. (1991). Homelessness and the Issue of Freedom. *UCLA Law Review* 39: 295–324.

Wallace, J. (1999). Deconstructing Gayatri. *The Times Higher Education Supplement,* July 30.

Wallace, S. (1965). *Skid Row as a Way of Life.* Totowa, NJ: The Bedminster Press.

Wilson, J. (1968). *Varieties of Police Behavior.* Cambridge: Harvard University Press.

Wilson, J. and G. Kelling (1982). Broken Windows: The Police and Neighborhood Safety. *The Atlantic Monthly,* March: 29–38.

Winerip, M. (1999). Bedlam on the Streets: Increasingly the Mentally Ill Have No Place to Go. *New York Times Magazine* May 23: 42–9, 56, 65–6, 70.

Wolch, J. and M. Dear (1996). *Malign Neglect: Homelessness in an American City*. San Francisco: Jossey-Bass.

4

Hiding the Target: Social Reproduction in the Privatized Urban Environment

Cindi Katz

The notion of a hidden city of social reproduction, suggests that the uneven relations and material social practices of social reproduction are respectively hidden and targeted by the neoliberal urban agenda fueled by 'globalization.' The hidden city is itself an outcome and a representation of what might be understood as 'postmodern geographical praxis,' but so too is the project of its unhiding. A discussion of the public–private Grand Central Partnership in New York City, reveals some of the ways that the agenda of simultaneously targeting and hiding social reproduction is pursued through preservation, and addresses how particular social actors and their activities are removed from view in the interests of ensuring 'orderly,' 'clean,' and 'safe' public space for others.

Postmodern Geographical Praxis

Postmodern geographical praxis calls to mind playful notions of heterotopia, reworkings of space by the forces of globalization and the effects of high technology, and the reconstitution of cities as global centers and microcosms of re-'worlded' neighborhoods. Most notions of heterotopia, which spring from a little corner of Foucault's work, play off the possibilities of engaging and embracing difference spatially (Foucault 1986, 1984; cf., Soja 1989). Like Lefebvre (1991) Foucault is clear that lived space is heterogeneous. How could it be otherwise – when people 'step outside themselves' they engage and indeed produce spaces of social relations and difference. But heterotopias are not so straightforward, it seems. They play off of 'utopias' in Foucault's musings. If utopias are literally no-places, heterotopias are 'real' spaces within "social spaces whose functions are different or even the opposite of others" (Foucault 1984: 252). Heterotopias juxtapose several spaces or sites that are 'incompatible,' and so 'function' either to "create a space of illusion that exposes" the partitioned spaces of everyday life as illusory, or to 'create a space that is other,' as ordered as our every-day spaces are 'jumbled' (Foucault 1986). If the former is the hetero-topia of 'illusion,' the latter is the heterotopia of 'compensation.' And in a revealing aside, he wonders whether 'the colonies' did not fulfill this latter function. According to this frame, prisons are a classic het-erotopia.

Why their appeal, I wonder. Except of course to those unmarked pro-ducers of space and social relations whose sense of order may be at stake. 'The colonies' and prisons produce ordered spaces that relieve particular social actors from the 'messy' and 'jumbled' world of differ-ence. It takes astonishing privilege to even imagine such transcendence, and great power to produce it spatially. This is to say nothing of the entitlement that could provoke the need to construct a concept which points out that the partitions of everyday life are illusory. Ask any 'working mother.' So, while the notion of heterotopia may sound appealing – a goal of spatial politics intent on producing spaces that engage difference and welcome diverse publics in all their heterogene-ity – it is an imagined and 'real' outcome of spatial practices of domi-nation and privilege. In the wake of decolonization, heterotopias of compensation have taken the form of 'fortress cities' and privatized theme-parked 'public' spaces, among others. These real spaces, as out-comes of postmodern geographical praxis, bear scrutiny in relation to the question of 'compensation' raised by Foucault.

Postmodern geographical praxis, as Giuseppe Dematteis (in chapter 5 below) suggests, is also seen in the fragmentation of territories into various 'functional units' that fasten onto global networks, leading them along paths which give them more in common with distant places than their neighbors. Indeed the fragmentation often hurls those unfastened neighboring territories further away from those global networks. And that is often the point. Uneven development, the quintessential spatial form and practice of capitalism, works at a variety of scales and its inter-nal combustions have latterly produced spatializations that simultane-ously draw some territories ever tighter into a global grid of power while pushing others off the map of the world that matters. As a territory gets bound closer to the heart and pulse of capitalist power, the relative dis-tance of its neighbors from those centers is that much further, to say nothing of those territories further afield. This compression-expansion expresses another of the reworkings of space that might be understood as postmodern geographical praxis. These spatial dynamics produce volatile 'cities of difference' whose fault lines warrant examination.

Among their manifestations are cities with intensified connections to global finance and production capital, and sharply differentiated residential neighborhoods lived in by an increasingly mobile population. There is, for instance, a 'first worlding' of selected cities in underdeveloped countries and at the same time 'third worlding at home' (Swyngedouw 1997; Koptiuch 1991). While these intertwined phenomena are a variant of the compression-expansion described above, they also reflect contemporary peculiarities in the geographic scales of production and reproduction. As capitalist production is increasingly globalized and mobile, its commitments to social reproduction in particular places become more tenuous. If, as Neil Smith (in chapter 7 below) suggests, the journey to work marked the separation of production and reproduction in the modern city and the scale of the city was defined by the limits of daily travels between home and work, the globalization of capitalist production, the relative cheapening of transportation, and the mobility of production sites have pushed that journey airborne and outwards for many. This shift alters both capital's commitments to social reproduction and people's strategies for ensuring its outcomes. The city is no longer the hearth for its own sustenance and reproduction. Future and present workers are readily (and more cheaply) reproduced elsewhere. The resulting spatializations – of fluidly shifting urban and regional forms, jostlings of class, race, nation, sexuality, and gender in incessantly reworked neighborhood and work spaces as well as at the scale of nation-states (the 'messy' realm for which Foucault's heterotopias provide relief?), and the excessing of particular populations whose reproduction no longer matters to capital accumulation – might be understood as yet another instance of postmodern geographical praxis.

John Berger's (1974) typically stunning insight that 'now' "it is space not time that hides consequences from us" was a startling recognition of 'simultaneity and extension' (cf. Soja 1989 who brought this insight to light for most geographers). Berger was writing about portrait painting and literature, but the timing of his statement seems significant. *The Look of Things* was written shortly after the unhinging of the Bretton Woods currency exchange system which enabled hard currencies to float in relation to one another rather than be pegged to the US dollar, itself recently taken off the gold standard. It was also around the time of the first oil shocks, and at the close of the most visually mediated war ever. The sociospatial processes associated with new forms of globalized capitalism and cultural relations had been set entrain by these events among others, and their consequences – obvious in hindsight – were astutely grasped by Berger as they began to take hold. Whether or not, as taken up by Soja (1989), this recognition is one of the hallmarks of post-

modern spatiality or a critical blow to historicism in favor of the spatialization of social theory, it is clear that the spatial forms associated
with increasingly globalized capitalist production are indeed masterful
at hiding the consequences and contradictions of the social relations
associated with it. These hidings are part and parcel of each iteration
of postmodern geographical praxis noted above.

Hiding Consequences

Given the brilliance of Berger's insight, it is interesting that there has
been so little uptake on it as a call to action or at least more nuanced
understanding rather than as a chillingly insightful observation. What
does it mean for space to hide consequences? What are the politics of
either living with or confronting this condition? If one of the tasks of
critical geography is to analyze the historical geographies of capitalist
development, it is important to examine the historical geographies of
contemporary capitalism as part of this project. Just as radical historians engage with standard histories to exhume forgotten and erased
histories and social actors, radical geographers confront the power of
geography to eclipse particular people and material social practices.
Such a project of unhiding offers a more positive spin on what might
be thought of as postmodern geographical praxis, and suggests its
urgency.
    I want to address two recursive instances of hiding; the geographies
that hide the consequences of shifts in social relations of production and
reproduction associated with globalized capitalist production; and the
analyses of space that focus on this globalized scale and thereby hide
(and miss) other scales and material social practices of significance; in
particular those associated with social reproduction. Before I tackle a
particular instance of hiding and unhiding in New York City, a word
about this endeavor as a postmodern geographical praxis. Part of the
critique of masculinist modernist thought has addressed its tendency to
privilege the visual, and more pointedly, the ways that privilege erases
the viewer from the scene and seen. 'His' gaze is unmarked, yet what
he sees has the effect of rendering or 'worlding' – if not quite producing – the world. But salutary as this insight was in terms of recognizing
the importance of position and calling for more enmeshed or differently
situated accounts apart from the panoptic, seeing and being seen maintain their potency. Understanding and 'exposing' the ways that things,
people, and social relations are made visible or invisible to the public
eye remains an important political project. Thus, I want to expose some
of the ways contemporary urban space is 'hiding the consequences' of

globalized capitalist production, and call attention to the ways most analyses of 'globalization' do little to unhide these consequences because their focus on finance, information flows, and production rehearses the panoptic gaze.

My focus is on material social practices of social reproduction hidden from public view in the reworked public environment of New York City. It traces the congeries of spatializations that I have suggested might be considered as postmodern geographical praxis; the compression–expansion of space that accompanies the fragmentation of territories, the simultaneous 'First Worlding' of some sites and 'Third Worlding' of others produced by migrations of people, influxes and outflows of capital, and the transnationalization of various cultural forms and practices; and finally in the face of these shifts and their messy spatializations, the production of a particular heterotopia at the heart of the city.

While 'globalization' is highly touted and over-assumed, there unquestionably has been an expansion of capitalist production globally such that components for single products are produced in diverse locations and assembled in yet another, and flexible production technologies are deployed to determine the location and specifics of production in any given cycle with much enhanced fluidity. This process gives capital enormous flexibility in shifting the actual grounds of production in response to increased labor or other costs, problems in quality, over- or under-production, the regulatory environment, or labor militancy, among other things. These shifts are driven by and in turn propel a number of changes in international finance, circulation, information flows, and the role of the state which have been well addressed in the literature and are not my concern here. While the effects of globalization are less totalizing – more partial – than is generally assumed, the globalization of production does alter capital's commitments to particular locations and unhinges many of the old assumptions concerning the relation between production and reproduction. As capitalists can and increasingly do draw on a globally produced workforce, they need not invest in social reproduction in the ways they formerly did (cf. Katz 1999). Not only is labor's ability to wrest these investments from capital weakened in the process, but so too is the ability of the nation-state or other geographic regions such as municipalities to do so. Instead they scramble to offer tax and other concessions to an increasingly fickle, fluid, and demanding production capital. These actions further reduce the resources available for social reproduction in particular locations of production.

These issues have undergirded and propelled the transformation of New York City over the past three decades. Practices such as 'the end of social welfare as we know it'; the privatization of an increasing array

of social goods, public services, and public spaces; and punitive antila-
bor practices reflect the unchecked power of corporate and state 'revan-
chism' to discipline and deny the aspirations of working-class people
(cf. Smith 1996; Piven 1999). But this power is enhanced by the very
ability of 'space to hide consequences'. Through differences of scale,
place, and territorialization – particular and conscious spatial strategies
– some of the contradictions of disinvestment, privatization, welfare
cutbacks, and the like are hidden from view. This hiding not only
emboldens those imposing these practices because some of their harsh-
est consequences are masked, but that very masking also reduces the
ability of would-be opponents to effectively confront these processes.
These interwoven practices build upon and intensify the uneven social
relations of production and reproduction under capitalism aided and
abetted by particular spatial strategies that exacerbate the unevenness
as they cover their own tracks. All of these issues come together strik-
ingly in Grand Central Station in the middle of Manhattan.

Fiscal Crisis and the Privatization of Public Space

For a variety of local and global reasons, the political economy of New
York City changed dramatically between the long boom following
the Second World War and the end of the Vietnam War in the middle
of the 1970s. These changes were manifested spatially in the decline and
virtual abandonment by capital of certain neighborhoods, suburban
development, and the deterioration of New York's infrastructure and
public environment. These shifts, which were fuelled in part by the
financial ricochet between suburban development, urban disinvestment,
and 'urban renewal' (predicated on 'slum clearance' and urban decon-
centration); the breaching of Bretton Woods, heightened capital fluidity, the oil shocks, and the increased globalism of capitalist production;
and the shift in New York's economy away from manufacturing and
port related industries to the finance, insurance, and real estate (FIRE)
industries, had dramatic repercussions on the relationship between production and reproduction in the city and its sociospatial manifestations.
Employment opportunities were increasingly bifurcated during this
period into high-compensation, high-education jobs and jobs that
required little training and offered little security, compensation, or benefits. As the Keynesian-backed promises of Fordism fizzled and New
York City was plunged into fiscal crisis during the middle of the 1970s
when capital fled to other sites of production and investment, New York
visibly suffered the effects of wholesale disinvestment in education,
housing, healthcare, social welfare, and the city's vaunted parks system.

Growing numbers of individuals and corporations left the city for obvious reasons, while those who could not afford to leave were marooned by capital and eventually the state, both of whom blamed them for their failures.

Even major corporations threatened and began to move their head-quarters away from the City. New York's cachet as a world city, eco-nomic magnet, and thriving cultural center faltered and seemed to teeter on the edge of imploding only a quarter of a century ago. The crisis abated when the city was bailed out by a ruling-class run public–private partnership – the Municipal Assistance Corporation – which issued 'big MAC' bonds to stabilize the city's finances while ensuring that banks and other financial institutions were not at risk. The Corporation instated conditions upon the debtor city that had noncoincidental simi-larities to the structural adjustments that were imposed by the Interna-tional Monetary Fund upon various debtor 'Third World' economies at around the same time. During this period and the years following it, investments flowed to private developments that furthered the ambitions of the wealthy people of New York while the public landscapes of social reproduction further deteriorated as pools of investment dried up, avail-able public funding shrivelled, and powerful constituencies turned away having secured these goods for themselves privately or outside of the city. Among the sites of deterioration were public schools, public housing, public transportation, and public spaces including parks and playgrounds. One of the most visible sites of deterioration was the Grand Central Terminal, one of two main railroad terminals in the city, which suffered from deferred maintenance, neglect, and its availability to (ware)house those evicted and excluded from other spaces such as workplaces, homes, mental institutions, and devastated neighborhoods.

The relationship between the visible city of production or circulation and the invisible city of social reproduction is direct. While many have focused on the flows of finance and production capital and the circula-tion of wealth and power at the scale of the municipality, region, or nation-state; or attended to the actions of ruling elites, corporations, financiers, and the state, I want to look more closely at what has become of social reproduction under these circumstances, and how its excesses have been spatialized. New York City's Grand Central Terminal is a case in point. The long neglected Beaux Arts rail terminal was lavishly ren-ovated over the last ten years through the aegis of the public–private Grand Central Partnership, a powerful midtown business group which early on set the standard for effecting a neoliberal sociospatial agenda. The result is a startling heterotopia, as gorgeous as it is demonic.

Through a self-tithing scheme sanctioned by the municipal govern-ment of New York, the Partnership, founded in 1988, created and

financed a large Business Improvement District (BID) centered on Grand
Central. As of 1998, property owners in the District paid annual assessments of 15.4 cents per square foot of property, giving the Partnership
an annual budget of $10 million, and a lot of power. Initially motivated
by the departure of the Mobil Oil headquarters from New York to
Houston, the leaders of the Grand Central Partnership were determined
to stanch the flow of capital and corporations from the city, and to
remove the tarnish – literal and figurative – from years of public disinvestment and neglect stemming from New York City's fiscal crisis in the
mid-1970s and the circumstances that led up to it. The perspective was
corporate, white, bourgeois, and masculinist. Mobil bade farewell to
New York with a videotape documenting the distastefulness of an executive's journey to work through Grand Central. The message was that
the residues of more than two decades of economic restructuring,
particularly the disinvestments in the city's solid manufacturing base,
accompanied by the white and middle-class flight that began in the
1950s but took serious hold in the 1960s affecting virtually all residential areas of New York, had become too much for the suburban
executive to stomach. Among the distasteful debris of these shifts
depicted in the tape were homelessness, panhandling, and depressing
infrastructural decay.

The Grand Central Partnership coalesced to deter further corporate
flight from Manhattan, and to revitalize midtown in ways that would
draw new forms of investment and bring middle- and upper-class visitors – local and not – back to New York. It is instructive that both the
motivation and the response of the BID involved and revolved around
the spatialization of political-economic and social problems. The BID's
evolution marks a turning point away from even a pretense of state-
supported social welfare intended to bring poor and 'disadvantaged'
people into the workforce and full consumership, to a privatized strategy that assists the wealthy and powerful to eliminate, escape, rise
above, and transcend the problems unsolvable by capitalist welfare 'as
we knew it.' In spatial terms the Grand Central Partnership produced
the very essence of heterotopia, and the strategy worked. How could it
not have, given the level of funding available to the Partnership and the
powerful affiliations of its leaders. It also worked because the Partnership's bid was embedded in the larger political economy of reinvestment
in New York by major financial institutions; the restructuring of global
capitalism in ways that produced and relied upon a few high-
technology financial markets or 'global cities,' among them New York;
and the emergence of neoliberalism, associated, among other things,
with massive public disinvestment in social spending and the
privatization of all manner of public and municipal services.

These trends and issues precipitate the very concerns with visibility and invisibility associated with power, wealth and poverty that are at the crux of my concerns. The Grand Central Partnership makes visible certain relations of wealth and power while explicitly hiding their undergirding and fallout. The stunning renovation of Grand Central Station as much as anything else in New York marks a landscape of power that both hides and hounds away the uneven relations and material social practices of social reproduction that enable and produce the contemporary urban landscape of 'globalized' capitalism (cf. Zukin 1991). The simultaneous revealing (of spatial forms) and hiding (of social relations) that has been their stock in trade has allowed the Partnership and its allies to revel in the success of a city recaptured for the wealthy with few noticeable qualms or objections. John Berger's (1974) prophecy has come home to roost. And so it behooves critics of neoliberalism to look outside the much-celebrated space of the new Grand Central to find consequences. One does not need to go far; the homeless shelter and soup kitchen at St. Agnes church two blocks away is part of the Grand Central Partnership, and a brief look at its work reveals an interesting and troubling relationship to the Business Improvement District.

Unhiding the Consequences of Privatization

Homeless people, especially visible ones, are antithetical to business improvement. In a late 1980s lecture by Daniel Biederman, President of the Grand Central Partnership, laying out the Partnership's mission, it was clear that homeless people were problematic insofar as they were objects – litter in the executive's path – and not as subjects: people who might have been made jobless as a result of the shift in the city's employment structure from manufacturing to the finance, insurance, and real estate industries during the previous two decades (cf. Fitch 1993); or who might have been displaced from their homes by any one of a number of events during the same period including landlord abandonment, the drying up of federal support for public or low income housing under Nixon, and steep increases in housing costs in many parts of the city; or who might have suffered from the effects of another of the manifestations of disinvestment in social reproduction, deinstitutionalization, which proceeded during the 1970s without the development of adequate appropriate neighborhood or household support services.

Homelessness was a problem in midtown because homeless people were depressing at best and menacing pests at worst. As a result, right from the start, the Partnership in association with the nearby St. Agnes School, sponsored a number of programs for homeless people. Its intent, more than anything else, was to make homeless people invisible to the bourgeois visitor to midtown – whether commuter, tourist, daytripper, or investor. The rights of homeless people to decent housing or employment or the rights of their children to a stable education were far from the concerns of the Partnership's outreach program. This interpretation of the Partnership's motives, which were crystal clear in both Biederman's presentation and in the publications and work of Jeffrey Grunberg, the director of their homeless programs, sadly was bolstered by the discovery that the Grand Central Partnership employed people to harass homeless people away from the District, beat homeless participants in the St. Agnes programs, and 'employed' homeless people at subminimum wage. The Partnership was successfully sued by the Partnership for the Homeless because it 'employed' homeless people from the St. Agnes shelter to do cleaning, maintenance, security, office, and 'homeless outreach' work in the terminal and elsewhere in the BID for wages between $1 to $1.50 an hour. It turned out that the 'outreach' work sanctioned and even encouraged the harassment and roughing up of homeless people in the area so that they would be spurred to either enter the Partnership's shelter system or leave the District. The Partnership claimed that these workers – who were not 'welfare workers'[1] – were participants in a social service program called 'Pathways to Employment.' The federal judge who ruled in the case determined not only that these people *were* already employed – some of them worked upwards of a total of 1,500 hours and many worked well more than 40 hours a week – but noted pointedly in the decision that Mr. Biederman earned $335,000 at the time. The Partnership was ordered to pay the 40 workers named in the suit back pay and overtime. The ruling could eventually apply to more than 100 other illegally employed homeless workers (Lambert 1998).

Grand Central Terminal as a Heterotopia of Compensation

Despite these deplorable practices and others, the Grand Central Partnership has been widely lauded by the city's elite and media for its 'visible' successes: the 'cleaning up' and reordering of midtown with such things as signage regulation, street furniture, taxi kiosks, and private sanitation and security services; and for the stunning renovation of Grand Central in which no expense has been spared or detail overlooked. The Partnership is relentlessly self-congratulatory and hopelessly unselfreflective. Its leaders do not seem able, and certainly are unwilling, to imagine that their values and visions for the future are not universal. When I pointed out to Biederman, after his lecture introducing the Partnership more than ten years ago, that it was problematic to have a wealthy, white, male commuter as the standard upon which the midtown changes were planned, he reassured me without irony or recognition of what I was suggesting, that the suburban executive was a sort of 'everyman.' Such is the enduring power of the 'panoptic gaze.'

But this presumption is untenable at the heart of a city that is among the most heterogeneous in the world. It is all the more troubling given the sharp and racialized class differences that characterize New York, many of which were at the heart of the disinvestments that led, among other things, to Grand Central's deterioration in the first place. These differences are masked and papered over by projects such the Grand Central Partnership, the painful costs of which are buried in the seeming tautology that what is seen is what has been made visible. That visible monumentality is built on rendering invisible those who are on the losing end of the great and growing divide between rich and poor in New York and elsewhere. Recovering and probing the lost traces or hidden landscapes within the monumental or formal built environment of Grand Central is an important countermeasure to the erasures of these people and their claims to the city.

Grand Central Station is iconic for me. It was the space of betweenness in which I felt myself independent, engaged with difference, and entered my political life. It is the only space about which I have recurring dreams. These are often about the hidden and liminal spaces of the terminal around the tracks and their portals. So my concerns about the space are deeply inflected with personal meaning. But Grand Central Station is also an everyday space that I traverse routinely. Until recently my journey to work involved traversing Grand Central Station, and each day I marvelled at its renewed grandeur, fumed at the amount of space that has been privatized for upscale commercial purposes, and puzzled over the obsession with origins I witnessed there as well as in other recent preservation projects. In the case of Grand Central, the fascination with origins resulted in the construction of a grand staircase on the east end of the vaulted main space that matches the western stairway leading to Vanderbilt Avenue. The second stairway was in the original plans for the terminal but never constructed. Why this obsession with being more original than the original, a kind of ur-authenticity that involves going back to 'true' origins? During my walks through the terminal I would muse that it has something to do with the deep anxiety 'everyman' must feel these days in the face of 'difference' as certain forms of privilege are no longer guaranteed with easy recourse to secure origins. But apart from this sardonic musing – which recalls the impulses behind heterotopias of compensation – I think of those steps, leading only to a fully commercialized mezzanine, as somehow emblematic. The steps make a grand statement and show the determination to 'preserve' something that never existed while insinuating a reverence for authenticity. Yet all the pains taken to render visible this formal feature – to make a bold statement concerning the authentic Grand Central Terminal – are mirrored in the countereffort to hide certain kinds of people who inhabited and worked in the space, in many cases for decades. Their trace has been removed as inauthentic; a part of the landscape not worth preserving. It was not just the homeless people who were removed from the renovated premises, but shoe shiners, buskers, pamphleteers, proselytizers, and many modest retailers. The buskers and pamphleteers who remain in the station must be vetted by Grand Central's management, and their regulated presence reminds of the increasingly controlled and surveilled nature of all public environments.

Akin to the anxiety about origins, concerns with middle-class safety and what seems to be an increasingly common presumption of a class right to a certain kind of unperturbed passage through the public environment have resulted in the removal of all kinds of people from Grand Central and its environs, suggesting that they have, at best, unequal rights to the city. Many of these people, among them the shoe shiners and retailers, earned a modest living in the station and caused no harm to others, but their presence did not seem to fit the new image for Grand Central which includes Michael Jordan's expensive steak house taking up a quarter of the mezzanine, a Godiva chocolate shop, and the redundant but inevitable Starbucks coffee stall. Their lot, like those of so many aspiring middle- and working-class people in contemporary New York, seems of no moment to the architects of the neoliberal city – witness the enduring assaults on the City University, the Giuliani administration's 1998 harassment of taxi drivers and restrictions on street vendors, and the searingly high rents for even the smallest commercial spaces in many parts of the city. Yet, the texture of the city – its very driving force and unique quality – will be lost if such groups of people have no place in it. If Grand Central becomes as much of an ordered contrivance as a Disney production[2] and its commercial attractions no different from any upscale mall, the Partnership's 'operation' may be considered a success but the patient will be dead.

These issues of place in all of its meanings are at the heart of the fight against neoliberalism and the projects of preservation and privatization associated with it. Projects such as the Grand Central Partnership offer those in power a 'heterotopia of compensation.' The new Grand Central is a space of extraordinary order and control that does not just 'compensate' for the 'jumbled' nature of the spaces of everyday life, but rests upon material social practices that criminalize and evict targeted others who disturb 'the order.' This order is imposed by a private group deploying the privileges of class, race, and gender upon a more broadly shared public space. With public–private partnerships increasingly underwriting large-scale and highly visible projects of preservation, it becomes harder to find the everyday traces of lives spent among the monuments, as so many are simply erased or removed from view. It is politically important to retrieve those traces and insist on their right to presence in the city and the public eye if for no other reason than that the gap between rich and poor, and practices of domination, are reinforced and excused by their erasure. If these costs of preservation and business improvement are not made apparent, and the uneven workings of the city – what I am calling the hidden city of social reproduction – brought to light, then the vengeful politics associated with neoliberalism will proceed unchecked (cf. Smith 1996).

Opposing Spatialized Neoliberalism

But so much remains to be done. Exposing the hidden city of social reproduction alone is not enough. These same populations are not simply hidden in contemporary politics or its physical landscapes, but are targeted. Poor, racialized, and otherwise marginalized people are scapegoated and made to bear responsibility for their disenfranchisement, impoverishment, and marginalization in the unseemly and ever more vicious politics of rapacious neoliberal capitalism. Cloaking themselves in the diffuse mantel of 'globalization,' contemporary capitalists have managed to gut many of the gains of unionized labor, make the notion of a living wage a receding dream for the majority of workers and would-be workers, and discipline the multiply constituted workforce with threats of 'the other' either within or elsewhere. This in a period of astonishing economic growth. These are spatialized practices, and many of them revolve around social reproduction. For instance, in the US, Canada, and much of Europe, gendered and racialized immigrant labor – cheapened by the social reproduction work done by their relatives remaining 'home' – can be relied upon to provide domestic labor and other work of social reproduction such as child and health-care that in effect transfer a substantial subsidy of social reproduction to capital through individual bourgeois households and other social institutions (cf. Rose 1993). On the other hand, another stratum of immigrant labor is commonly counted upon to provide a skilled and well-educated workforce to northern industrialized countries that is produced outside their domestic schools, while others – marooned in those schools – are denied employment as inadequately trained. In these ways and others, the disinvestment in and neglect of social reproduction and the diminution of the social wage can continue without threatening capital accumulation or an increasingly globalized bourgeoisie.

Such transnational transfers of wealth from poor to rich notwithstanding, the denial of employment or a living wage has gone hand in hand for many in the US with a shrill rhetoric of victim blaming raised to national policy, and a harsh demonization of those in need. Not just the reality but the word welfare is under erasure. Taking the discursive turn even more seriously than ardent poststructuralists, 1990s reactionary bureaucrats and mean-spirited politicians have renamed welfare offices "employment" or "job centers," case workers in some states have become "financial planners," and Departments of Public Welfare have changed to "Departments of Transitional Assistance," "Work and Gain Economic Self-Sufficiency Programs" (i.e., WAGES), or "Family Independence Administrations" (Swarns 1998). Nominal politics replaces service provision as money is drained from the provision of permanent jobs, adequate transportation, childcare, or training that could make the transition from welfare to work more than just a turn of phrase. Indeed, a draft memo included in the training manuals of 'financial planners' in New York City's 'Family Independence Administration' called for a tactic reminiscent of the 'homeless outreach' recently advocated by the Grand Central Partnership. The memo indicated that employment was a 'secondary goal' of the job centers; the 'primary goal' was to discourage would-be applicants from requesting public assistance (Swarns 1998: 5). The Commissioner of the City's Human Resources Administration, Jason Turner, speaking of its welfare-to-work program, actually said in a television interview, "Work makes you free." When New York City and Auschwitz are working under the same banner, we are in more trouble than we think.[3]

Frances Fox Piven (1999) traces this state of affairs to many factors: the loss of jobs in recent decades as a result of technological innovation; the flight of production from traditional centers; and certain manifestations of that specter called 'globalization.' Piven's work demonstrates how finance capital has pinned various nation-states to the wall via the threat of capital flight, disinvestment, economic restructuring, programs for disciplining 'unruly' workers, all accompanied by a predatory scavenge for highest returns no matter what the social costs. The corporate commitment to any particular place or indeed to the workings of a democratic civil society stretch only so far as the commitment to profit. In the US this predation plays well with a cultural politics of individual self-sufficiency, contempt for poor people (despite their romanticization in our various myths of nationhood), a deep and easily fuelled racism, and a startling moral starchiness.

But Piven's key and compelling argument is that it is the expanded political ambitions of business in the past two decades that have driven the contemporary antiwelfare, antiworker, antipoor, antiwomen, antichildren, antiminority politics that have got us all reeling. These have ushered in an historically unprecedented and growing gap between rich and poor, drastic declines of real wages and jobs, the stripping away of benefits, safety nets, and services, and the defunding of social reproduction except the carceral for all kinds of 'excessed' populations. These too are written in the landscape. The city of social reproduction is wildly uneven these days and its effects are as palpable as they are distressing. Schools are in sorry states while prison construction forges ahead. Neoliberal public–private partnerships such as New York's Grand Central Partnership and the Central Park Conservancy, lovingly and lavishly restore, refurbish, preserve, and build those landscapes visible to and treasured by the bourgeoisie. The sanitized visible environment creates a sense of well-being and civic pride for those who count, while public housing decays and goes unbuilt, schools and schoolyards get more crowded and dilapidated, long-tended community gardens in gentrifying neighborhoods are 'condemned' (confiscated) for luxury or market-rate housing development, and parks and playgrounds in poor neighborhoods are allowed to languish; broken-down and unsafe (cf. Katz 1995, 1997, 1998).

Poverty, the oppressions of class, race, gender, and other forms of marginalization, and the people who bear these injuries – all are less visible in the public political discourse than at any time in last three decades. As the discussion of Grand Central suggested, those with power and money have engineered things so they (and their traveling selves, tourists) do not see poor people, dirt, dilapidation and the painful effects of the wholesale disinvestment in social reproduction. These painful reminders of the unevenness and fragmentation brought about by capitalism have been pushed out of the central spaces of the city, and significant rhetorical and physical vigilance is mounted against their return. In New York and elsewhere, the outer city has become a hidden city of social reproduction – its energetic social life as well as its tawdry failures have to be made visible both *in situ* and in the centers of power themselves. That visibility will come most resolutely from movements of and by the people who most strongly feel the pain of exclusion. Taking a cue from the fragmentation of territories that Dematteis (in chapter 5) associates with postmodernity and the process of compression–expansion within a globalized sphere of production that accompanies it, these movements should be imagined and built transnationally and across scale.

And this brings me to my perhaps perversely optimistic conclusion. Just as Dolores Hayden's work in the *Power of Place* brings to light those eclipsed and disempowered in ways that are quite empowering, I want to affirm that in our own times we – feminists, leftists, antiracist activists, and others – have disturbed the comfortable illusions of a beneficent globalized capitalism. We can draw strength from that history to encourage and help respark a broad based activist response to current conditions in the hidden city of social reproduction and more widely. The recent uprisings against the International Monetary Fund, the World Bank, and the World Trade Organization are a case in point. But we need to follow all the filaments of oppression and exploitation to connect issues of social reproduction with those of a globalized workplace and of environmental degradation, each produced and policed by northern-dominated financial institutions that serve corporate capitalism on a world scale.

While Piven (1999) argues convincingly that the political onslaught and the "crazed scapegoating" that accompanies it, result from the expansion of political ambitions among the business classes in the last twenty years, she has suggested that this happened during this period because they *could* do it. I agree, but let us also turn the answer around. Today's right-wing onslaught in the US (and elsewhere) is very much a response to – revenge for – the very real successes of the political mobilizations of the 1960s and early 1970s. The antiwar, civil rights, women's, labor, and anti-imperialist movements, working in concert, separately, sometimes even in opposition, fundamentally shook the stability of US ruling power in ways it had not been shaken before in the twentieth century. For many reasons these movements did not live up to their promise; apart from anything else, they were violently attacked and repressed. The expansion of ruling political ambitions in the 1990s is a continuation of that backlash and projects such as the Grand Central Partnership are its sugar-coating.

Perhaps it is cold comfort, but it is also empowering, to realize that our own political victories lie somewhere underneath the contemporary landscape of meanness and revenge. In revealing these victories (another project of unhiding) and connecting them to contemporary struggles for welfare rights, decent public education, housing, job security, public space, and the rights of prisoners, we may find the power, strength, and inspiration to again mount and win the fights for alternative landscapes of production and reproduction. An appropriate postmodern geographical praxis under these circumstances would not only work to reveal the consequences hidden by the fragmented, heterotopic, and rescaled spaces of postmodernity and contemporary capitalism – the hidden city of social reproduction, for instance – but would provide guideposts for seeing these spaces from different positions, and join in struggles to transform them into social environments where constituent and mobile publics could imagine and produce a livable world for themselves and their children. This project will be a lot harder than constructing a grand staircase, but it is sure to lead someplace far more important than a commercial mezzanine.

NOTES

Portions of this chapter were previously published as "Excavating the Hidden City of Social Reproduction: A Commentary," in *City & Society*, Annual Review, 1998: 37–46. These sections are reprinted by the permission of the American Anthropological Association. Not for further reproduction.

1   The term 'welfare worker' refers to a person receiving government welfare assistance such as cash payments, food stamps, or housing vouchers which now come with strict time limits that vary by state, and that require all able-bodied recipients (broadly defined), including those with young children, to work for their assistance. The hourly compensation for this work is well below minimum wage rates.
2   Daniel Biederman noted in conversation that Disney was a model for the BID. When I suggested that a little grime was part of New York's particular charm he shuddered in fussy disgust.
3   That Turner said this on television completely unaware of the statement's Nazi origins does not excuse it. Rather, it makes it even more frightening, for it suggests a common reservoir of sentiments towards and strategies for marginalized people in contemporary 'revanchist' New York and Nazi Germany (cf. Smith 1996).

REFERENCES

Berger, J. (1974). *The Look of Things*. New York: Viking.
Fitch, R. (1993). *The Assassination of New York*. London: Verso.
Foucault, M. (1984). Space, Knowledge, and Power. In *The Foucault Reader*, ed. P. Rabinow. New York: Pantheon, 239–56.
Foucault, M. (1986). Of Other Spaces. *Diacritics* 16: 22–7.
Hayden, D. (1995). *The Power of Place: Urban Landscapes as Public History*. Cambridge: MIT Press.
Katz, C. (1995). *Power, Space and Terror: Social Reproduction and the Public Environment*. Paper presented at the Landscape Architecture, Social Ideology and Change Conference, Harvard University, Graduate School of Design. (Available from author upon request.)

Katz, C. (1997). Disintegrating Developments: Global Economic Restructuring and the Eroding Ecologies of Youth. In *Cool Places: Geographies of Youth Cultures*, eds. T. Skelton and G. Valentine. New York: Routledge, 130–44.

Katz, C. (1998). Whose Nature, Whose Culture? Private Productions of Space and the 'Preservation of Nature.' In *Remaking Reality: Nature at the Millenium*, eds. B. Braun and N. Castree. New York: Routledge, 46–63.

Katz, C. (1999). Stuck in Place: The Local Consequences of the Globalization of Social Reproduction. MS under review for *Antipode*.

Koptiuch, K. (1991). Third-Worlding at Home. *Social Text* 32: 87–99.

Lambert, B. (1998). Two Business Districts Violated Laws on Wages, Judge Rules. *New York Times*, March 20: B1, B8.

Lefebvre, H. (1991). *The Production of Space*. Oxford: Blackwell.

Piven, F. F. (1999). Welfare Reform and the Economic and Cultural Reconstruction of Low Wage Labor Markets. *City & Society*, Annual Review: 21–36.

Rose, D. (1993). Local Childcare Strategies in Montréal, Québec: The Mediations of State Policies, Class and Ethnicity in the Life Courses of Families with Young Children. In *Full Circles: Geographies of Women over the Life Course*, eds. C. Katz and J. Monk. New York: Routledge, 188–207.

Smith, N. (1996). *The New Urban Frontier: Gentrification and the Revanchist City*. New York: Routledge.

Soja, E. W. (1989). *Postmodern Geographies: The Reassertion of Space in Critical Social Theory*. London: Verso.

Swarns, R. L. (1998). Welfare As We Know It goes Incognito. *The New York Times, Week in Review* July 5: 1, 5.

Swyngedouw, E. A. (1997). The Mammon Quest: Glocalisation, Interspatial Competition and the Monetary Order: The Construction of New Scales. In *Cities and Regions in the New Europe*, eds. M. Dunford and G. Kafkalas. London: Belhaven Press, 39–67.

Zukin, S. (1991). *Landscapes of Power*. Berkeley: University of California Press.

Scales . . .

5

Shifting Cities

Giuseppe Dematteis

Introduction

In a world dominated and controlled by networks of interaction and global flows, many of the assumptions upon which the territorial conception of cities was founded are increasingly fading away. The emergence of a variety of novel political-economic and technological conditions are, similarly, forcing us to reconceptualize territories and cities in relatively new ways. Yet perhaps the greatest difference, the key break with the past, lies with the fact that while previously the city was thought of as a primary, taken-for-granted entity – stable in time – it can now only be envisioned as *one* possible, deliberate construction: a local geographical order born out of the turbulence of global flows and with which it must interact in order to continue to exist.

It is this basic assumption which will form the focus of the first section of this chapter. Section 2 will query the consequences of glob- alization on the spatial form of the city, on its representations and on interpretations of urban territoriality. An in-depth examination of these issues in the third section will allow me to advance the thesis that a postmodern geographical praxis necessitates a shift from a physicalist (causal) conception of space, typical of late nineteenth-century geogra- phy, to a conception of space as a logical-symbolic operator, suitable for interfacing creative and analytical thought. In the final section of the chapter, I will discuss the conditions for a postmodern geography able to represent and to plan for urban complexity, concluding that although the necessary epistemological conditions may exist, the greatest diffi- culties remain of a political nature. Not only; since the current hier- archical structure of power imposes (on a global scale) performative geographical representations of a hypermodern type, a true postmodern geographical praxis is conceivable only in terms of resistance and conflict.

The Shift in the Urban Scene

In the first half of this century, the notion of global networks was essentially absent with discussion focusing, rather, on international relations and their cross-border traffic of people, goods, money, and information. Some of these phenomena were already 'global,' in the sense that they produced effects on a planetary scale, but this globality was limited and indirect. International networks were, in fact, constituted by preexisting territorial units: primarily nation-states, though also cities as well as other territorial subdivisions of the state structure. The public and private actors shaping then-transnational relations were rooted in these territorial bodies and this embeddedness was, in fact, a fundamental condition for the development of transnational operations. Networks were thus founded on and depended upon their respective territories and boundaries within which – and through which – state, regional, and local structures were controlled.

As long as the control and management of flows remained territorial, the networks could be governed in territorial fashion, even if with significant power differentials between strong and weak cities, small and large towns, rich and poor regions. When this was no longer possible, above all because of the increased mobility of capital and information (which enabled the reorganization of production and markets worldwide), the situation changed drastically. Relationships of interaction were established between increasingly autonomous networks and less and less sovereign territories, as neither was able to control the other.

With states and other territorial entities no longer controlling the incoming and outgoing 'immaterial' flows, the networks of these flows and of the transnational organizations that managed them would soon become deterritorialized and sovereign global entities themselves. Yet the global networks and their 'nodes' (companies, research centers, major financial corporations, etc.) could not operate without bonding with certain territories and cities. In fact, as theorists of flexible specialization, of the *milieux innovateurs*, and of comparative advantage have demonstrated, the outcomes of global competition continue to depend upon such bonds, even though these may be less stable than past notions of 'embeddedness.'

It should be noted, however (as critics of globalization have pointed out), that this continued dependence of networks on specific territories is valid only in general terms: only in select cases, such as that of the 'global cities,' can the location of the "nodes" of global networks (for example, major transnational corporations) in a given region or locality be considered necessary. It is much more common, in fact, for a large number of territories, all endowed with similar comparative advantages, to compete with each other to attract the same global investor.

The expansion of global networks has had substantial effects on the regional structuring (and destructuring) of territories. As local actors 'link into' global networks (to import and export goods, to attract investments, for the purposes of cultural exchange and promotion, etc.), they increasingly tend to make themselves functionally independent of the territorial entities of which they are formally a part. Today, for example, cities like Lille, Glasgow, or Barcelona operate as actors on the international scene without recourse to the mediation of their national governments – and sometimes even against the general interests represented by these same. This phenomenon is also observable at the microregional scale: when, for instance, a particularly enterprising neighborhood manages to obtain European structural funds to finance an urban recovery or improvement project, or in the case of a suburban town that offers transnational capital the best conditions for the location of a shopping center, a factory, a leisure park, and so on.

The overall result is a *fragmentation of territories* into autonomous functional units of various sizes, certain of these with explicit aspirations to political and administrative autonomy. In fact, each territorial fragment (city, region, neighborhood, small town, or aggregation of towns) which has successfully entered into global networks will likely follow a rather independent development path, bringing it increasingly 'closer' to other faraway places linked into the same global network, while 'distancing' it (by differentiating and disconnecting) both from contiguous portions of territory that do not host 'nodes' of global networks (or that host 'nodes' operating in different sectors) as well as from the territorial entities of a higher level to which they belong in functional or institutional terms.

This fragmentation, which heightens differences and imbalances at the microterritorial scale, also often translates into social fragmentation, in that within the various 'regions' it increases the divide between those social classes that are linked directly or indirectly into the global networks – and those excluded from them in terms of income, substantial rights, and opportunities. One result of this is the social polarization evident in many great metropolises; the source of conflicts apparently without solution.

This shift's geographical consequences are numerous. Above all, the global networks, now largely independent of the individual territories that generated them in years past and upon which they depended until around the 1960s, are redrawing the planet's regional patterns and urban hierarchies. The sizes, geometries, characters, and dynamics of regional and urban spaces now derive largely from the interaction of the local systems with the networks that intersect them; networks which concentrate their operational and decision-making 'nodes' in a few dense areas such as major cities or metropolitan areas; that remain, for the time being, the only fixed points of a variable and relatively unstable territorial pattern.

Among them, it is the global cities that dominate. More than individual entities, however, such cities should be seen as a web of global networks through which the latter try to exercise social control on a planetary scale. The emergent urban hierarchy is no longer Christallerian, no longer based on territorial range (Conti, Dematteis, and Emanuel 1995). It is, rather, a hierarchy of networks, within which the relations between 'nodes' (the cities) not only do not depend on distance but are also, in fact, to a certain degree *interactive*. Local systems, as bearers of specific comparative advantages, can thus condition the choices of global organizations and resist their domination. If, then, global cities and networks (with their growing concentration of strong powers) can be envisioned as moving in a hypermodern direction, their interactive relationship with the local urban systems has a character of complexity that evokes a possible postmodern scenario.

The Shift in the Spatial Form of the City and the New Urban Territoriality

The twentieth century has witnessed frequent adaptations in the definition of the city to the accelerated variations in its form. Premodern *urbs–civitas* spatial coincidence models – still present in Weberian formulations – were followed by conurbation images, conceptualizations of the city-region and the urban field and, most recently, the notion of edge cities (Garreau 1991) – a spatial networking continuum around the scattered fragments of an exploded centrality (of which Los Angeles is certainly a prototype (Soja 1989)). Today's cities, in fact, can no longer be seen as stable, individual entities, made up of a physical 'body' (*urbs*) and an organizational 'mind' (*civitas*) which allows them to formulate strategic decisions and actions. This capacity, which in the past was considered a 'natural' attribute of cities, has now become something to be 'constructed.'

The recent changes in the spatial form of urban systems can be considered, in part, as simply the quantitative extension of previous suburbanization processes, albeit with a wider decentralization of residential patterns and related services and infrastructures. Yet, in many ways, the recent urban expansion differs in significant fashion from simple urban sprawl or from the extension and merging of contiguous urban fringes. It is characterized, in fact, by a number of qualitative changes such as spatial fragmentation, less regular center–periphery gradients, the weakening of spatial hierarchies in outer suburban areas, and the rise of new centrality patterns (Berry and Kim 1993).

Traditional terms such as 'suburb,' 'rurban,' 'sprawl,' 'rural-urban fringe,' etc. are thus no longer appropriate to describe the new forms of outer urbanization, and even the classical concepts of the 'daily urban system,' 'metropolitan area,' 'city region,' etc. seem hard pressed to capture the complexity of the new urban systems. At the theoretical level, the challenge lies with the conceptualization of complex urban systems at various levels of their spatial organization: from local networks of individual actors linked by proximity relations to the networks of cities interacting in a space free from distance constraints.

The new functional, social, and physical forms of urbanization have been described in years past with concepts such as the French notion of *périurbanisation* or the American *urban field*: the former referring to the outer expansion of the city, the latter to the whole extended urban system resulting from it. The essence of the *périurbanisation* concept was perhaps best articulated by J. B. Racine (1967) as a form of urbanization separated from the traditional form of the city; current interpretations have focused, however, on the production of new urban spaces which do not directly depend on a random diffusive decentralization from a central core. In Italy, the *città diffusa* concept has been adopted to indicate the overcoming of the city as a simple nuclear or areal entity and the evolution from the sprawl of settlement patterns to a complex reticular model combining physical grid expansion and the spatial redistribution of specialized functions (Indovina 1990; Boeri, Lanzani, and Marini 1993).

At the scale of the whole urban system, we can find a similar idea echoed in Friedmann's concept of the *urban field* (Friedmann and Miller 1965; Friedmann 1978): a large territorial unit evolving towards a multicentered networked spatial pattern, in which the former core (the 'mother city') is losing its traditional dominance to become one of the specialized centers of the unit as a whole. At the national and continental scale, we can conceive of multiple urban fields which, increasingly, tend to engulf the entirety of the territory, thus eliminating non-urban space, the urban frontier, and, in the future, the possibility of any further spatial extension of urban units.

The analysis and governance of urban systems now articulated as new forms of continuous settlement networks is certainly a major theoretical and political challenge. For one, the urban field model that best anticipated many of the functional and social aspects of new urban structures, now appears ill-suited to grasp some of the essential features of these same. We should recall that the main political issue in the new processes of urbanization highlighted by Friedmann's analysis was the avoidance of 'collective disaster' resulting from an urban form determined by unrestrained consumer choice alone. Freidmann's proposed solution, in fact, was to turn the urban field into a regional urban community ('a true *civitas*') with a tiered hierarchy of four multipurpose government levels: neighborhood, district, regional (urban field), state. Yet the new urban structures are no longer characterized by evident, self-contained boundaries, thus undermining the rationale of Friedmann's model: the possibility of identifying a given territorial basis for the definition of extended urban communities as new forms of *civitas*.

The new extended urban systems seem, rather, to be a mesh of different spatial structures *à géometrie variable*, shaped by flows and connecting individual territorial units at different scales, both through proximity links as well as through long-distance ones. Simply put, the traditional *civitas* model, based on a taken-for-granted organic territorial social cohesion, no longer 'fits' the complex and extended social forms of today's urbanization. Even where the spread of the city maintains its distinctiveness, it is highly doubtful that the individual and collective actors that make it up constitute a cohesive group just because they live in the same place; in reality, each of these latter may hold a functional and identity membership in a number of supralocal networks of businesses, institutions, associations, and cooperative bodies. And it is these (most often global) networks which run through cities and connect their actors across distances, weakening the traditional bonds of internal cohesion founded on physical proximity.

Yet although the city-actor rooted in territory is no longer an *a priori* certainty, recent typological analyses (Cattan et al. 1994; Bagnasco and Le Galès 1997) do stress that many cities still continue (and perhaps now even more than before) to behave as collective actors in the now-global spaces of competition and cooperation. Recent transformations have not succeeded, therefore, in eliminating urban territoriality but have, rather, modified its nature: accentuating its role, shifting it from passive to dynamic, from the simple result of a long historical process to the product of local self-organization, from a utility value to be made use of in a limited geographical area to a sort of exchange value, from an 'asset' to be conserved to one of risk capital to be ventured in global competition.

As the stratified deposit of a natural and cultural heritage, the physical city and its territory no longer have a value in and of themselves but, rather, for what they *become* in the valorization processes. The city as a local society is also no longer identifiable from its stable rooting in a given territorial milieu. It is, instead, a changeable, connective combination of actors, each operating as a 'node' of local and global networks: it is thanks to these actors that such networks come together, interact, and interconnect within the city. And it is for this reason that, in the virtual space of networks and flows, cities *continue* to exist as territorial 'bodies' and as social aggregations that play an essential role in the processes of globalization that seem destined to destroy their identities.

An important shift has occurred, however. The urban territorial milieu no longer offers the global network true *rooting* but, rather, mere *anchorages* (Weltz 1996). An urban milieu can therefore be defined or described only through the *effects* that it produces. It is only thus that we can explain the current leading role of cities, their character as attractors-connectors of global networks, as well as the formation of local social networks around projects geared to enhance the conditions and resources of the local context (Governa 1997).

The removal of obstacles and friction in the worldwide deployment of market forces does not, therefore, create *homogeneity* between territorial systems but, simply, multiplies the *connections* between them (Hannerz 1996). Territorial systems compete to seize these connections – in other words, to attract the 'nodes' of the global networks made up of the major public and private transnational organizations, financial markets and specialized services, scientific and technological cooperation, mass media, etc.

Globalization has thus transformed the conditions and resources specific[1] to the various urban and regional urban milieux into *competitive advantages*: it has made them emerge as powerful factors of local diversification. Within each city, local actors are thus induced to organize themselves, to forge new bonds so that the city becomes an attractor of global networks. For their part, such networks envision the anchorage of their 'nodes' in diversified local systems as a set of opportunities, capable of increasing their competitiveness on the global scale both for the possibility of exploiting the particular local resources and conditions in the multilocated and flexible organization of production, as well as ensuring access to differentiated segments of demand, with return effects on product and organizational innovation. The hypermobile global economy thus works thanks to a wide range of resources which are deeply embedded in places (Sassen 1998).

The competition between cities engenders new forms of territoriality. If by territoriality we mean the symbolic, cognitive, and practical mediation that the material aspects of places and their milieux exercise on social action (Raffestin 1981; Turco 1988), we should also distinguish between two of its forms. The traditional form, which we can define as *passive* or negative territoriality, concerns the demarcation and control of the territory (Sack 1986; Agnew 1987) and is also common to animal communities. The territoriality now stimulated by global competition is, instead, *active*, in that it consists in enhancing the conditions and potential resources of the various milieux in local development processes, i.e. by activating positive-sum games.

In fact, it is by acting on these 'territorialities' that local urban systems interact with the global system. Yet, as we have noted, such ungoverned interactions can produce territorial fragmentation and social polarization. Their 'governance' implies policies aimed at broadening the network participation of local actors and connecting together the various local urban systems through networks of cooperation. A truly postmodern city thus appears as a voluntary construction of the *civitas* in the face of the disappearance of the traditional *forma urbis* and the increasing tendency of global competition to fragment territories and societies. Such a construction must, necessarily, involve both local social networks and supralocal urban networks; it must create networks of cooperation without, however, expecting to eliminate competition or conflict.

The Shift in Geographical Images: From Causality to Metaphors

Such shifts in the spatial form also go along with accordant shifts in the geographical images of the city – and thus changes in its very meaning. There is a growing gap, however, between the new spatial form and current representations of urban processes. In fact, the paradigms of neoclassical (late nineteenth-century) geography, founded on causal relationships of physical proximity, are no longer capable of representing change due to long-distance virtual interactions. What is needed, then, is a new geographical praxis able to represent current hyperconnected, hypermodern urbanization and to plan for truly postmodern urban conditions.

A first step in this direction is to recognize that 'geographical space' is not a real entity, but a simple logical operator. In this way, what is commonly thought of as geographical (or 'spatial') causality becomes a simple semiological relationship, enabling us to put together a great variety of phenomena within a single metaphorical representation: in a sense, a weak version of the old notion of 'geographical synthesis.'

In the opening pages of *Economy and Society*, Max Weber notes that: "there are many different ways in which to try to define the city. All cities have in common just one thing: that each one is always a circumscribed settlement, at least relatively; it is an agglomeration, not one or more isolated buildings." For Marx and Engels, writing in *The German Ideology*, "the city is a concentration of population, means of production, capital, pleasure and needs, while the countryside makes the opposite emerge: isolation and separation." We could go on endlessly with similar citations; suffice it to say, however, that even those who do not conceive of the urban phenomenon as a geographical fact still adopt a conceptual category – the city – whose immediate referent is a physical and spatial entity. A geographical image is thus used to represent a set of facts that are of a historical, cultural, political, social, and economic nature.

It does not appear that such a dual meaning can be easily eliminated. The geographical concept of the city (its *immediate* or *literal* meaning) is certainly not the principal one. As a mental image it plays, in a certain way, the role of the 'signifier' with regard to the mediated or metaphorical signified (the city as a social fact) which is, certainly, its key connotation. But when we adopt the term 'city,' even if we only mean a set of social relations, we always and inevitably indicate, at the same time, a physical and spatial reality. A geographical distinction is thus made between urban and non-urban space; a correspondence established between the city sign and certain aspects of perceived reality.

The scarce attention paid to this semantic aspect has two principal consequences. The first is to confuse and conflate the two meanings of city, as though they were equivalent – while, as we have seen, the operation consists in representing an essentially social fact through *some* of its physical and spatial attributes. The second is to thus implicitly suggest that the relationship between the physical city and the urban social community is of a causal nature. In this way, a *semantic* relationship is transformed into an *ontological* relationship. What could, at most, be conceived as a hypothesis awaiting verification, is assumed as an evident fact. We should note here, however, that there are numerous other conceptual categories within which such a conflation also occurs: country, nation, center, periphery, marginality, etc. All categories endowed with a primarily geographical meaning that, however, recalls another, social, one and that, at the same time, denote certain physical and spatial entities that can be found in observed and observable reality.

In human geography (and in the social sciences more broadly), we often adopt metaphors of physical-spatial meanings in order to indicate phenomena of particular complexity, the field of which is neither described nor interpreted by a unified and universally accepted theory.

In such cases, mental images are adopted to depict concrete situations as a whole within which different facts are represented together in all their complexity.[2]

Geographical practice can thus give an important cognitive contribution if its descriptive categories can be linked to certain theoretical formulations. What is needed, then, are conceptual categories capable – through metaphorical inference – of systematically connecting geographical descriptions to various theoretical frameworks and thus of contributing to extending their experimental interpretation. This implies, of course, that we must give up the apparent scientificity of causal models based within pseudo-theories of objective geographical space in order to dedicate ourselves to the critical elaboration of the semantic and syntactical contents of geographical representation and to setting up the linguistic procedures necessary for the selection and communication of theoretically important empirical information.

The geographical practice that derives from this may be termed 'deconstructive,' in as much as it is critical of 'normal' geographical images – at the same time, however, it is also 'constructive' of representations that knowingly adopt the vast potential of analogical-metaphorical discourse (Dematteis 1997). If we limit ourselves to but decoding the unconscious metaphorical images of neoclassical geography – to then translate these same into the abstract conceptual categories of the social sciences – we deprive ourselves of the formidable heuristic tool of geography as a metaphorical interface between empirical observation and theory. In other words: playing by the rules of normal geography, though keeping firmly in mind that it is, in fact, a game, i.e. that the spaces we talk about are not the independent variables of a causal relationship, but *metaphors* which, by drawing upon the complexity of the world, allow for the falsification of theories – and thus their enrichment and critique.

A knowing use of geographical metaphors is not limited to the cognitive framework alone, however; it also speaks directly to very 'practical' considerations. Social relations between people, after all, take place through relationships with 'things,' so that (especially in human geography) the correspondence between physical-spatial relations and social relations is not purely metaphorical but is also, to a certain extent, quite real. This 'concreteness' of geographical images is of decisive importance at the pragmatic level. As I have tried to illustrate elsewhere (Dematteis 1985), the spatial metaphor can function as a self-realizing prediction. Geography thus represents (existing or possible) social relations through the 'evidence' of a reality observable by one and all. And since it is difficult if not impossible to refute what is represented with such evidence, geographical images are extremely persuasive, convincing us to adopt forms of social behavior which conform to the realities represented.

A Geographical Praxis to Deal with Urban Complexity

I believe that it makes sense to talk about the postmodern as, above all, a *method* (Dear 1986; Minca and Dear 1997). To my mind, it consists in not reducing complexity; a reduction which usually hinders us from dealing with 'reality' and its problems in adequate fashion. This is certainly the case of the city: today, an increasingly complex reality emerging from the interactions between deterritorialized global networks and local territorial systems, each operating according to different principles of order not reducible to a single higher rationality.

At the representational level, there persists the problem, however, of putting together two apparently contrasting spatial logics (though which are, in reality, inseparable): that of the 'horizontal relations' (or spatial interactions) which govern the virtual networks, and that of the 'vertical relations' which express the relations between local societies and their specific territorial milieux. The vision of space as an objective entity, typical of neoclassical geography, transforms this logical opposition into an *ontological exclusion* and resolves it by *separating* the two types of relations. Taken to its most extreme consequences, this simplifying trend produces two conflicting images of the world. One, made up exclusively of pure horizontal relations, writes the world as a network of interactions and flows not linked to particular places; disregarding any utility values or local languages that cannot be translated into exchange values or universal languages (Raffestin 1981). Such a geography can be termed hypermodern as it is particularly suited to representing the most evolved stages (and, perhaps, the very essence) of modern capitalism and the market economy as a system of free 'horizontal' relationships (Dematteis 1994). The second vision, on the other hand, eliminates horizontal relations at the expense of vertical ones, making everything dependent upon the specific features of places. Under the scientific guise of a pseudo-environmental determinism, it constitutes what is, at base, a nostalgic return to a premodern geography, typical of those precapitalist realities in which the local tends to represent particular utility values in the absence of a system of exchange based upon general equivalencies.

Yet if, instead, we consider networks and territories as *metaphors* which represent social relations, which represent relations with local milieux, then horizontal and vertical relations become simply a means of representing a *single* reality – viewed simultaneously from diverse viewpoints, and maintaining its intrinsic complexity. The geographer can thus assume a critical and active attitude adopting, time after time, the most suitable means of constructing representations of this 'reality,' observing from varied perspectives, taking on a both global and local vision.[3]

Briefly, then: *premodern geographies* tend towards silence, embodying local meaning in its pure state; *hypermodern geographies* tend towards noise as global signifiers free of signifieds. The 'intermediate' modern phase appears to be a local which expands to encompass all other 'locals.' *Metaphorical geography*, on the other hand, represents the global as a plurality of 'locals': as its active, autonomous, and necessary components, with distinct properties and languages, not reducible to those of the global system. It thus maintains the possibility of connecting general signifiers to varied local signifieds – and, at the same time, of connecting such local signifieds to each other, with the global represented as a network of locals. We can term such a geography a *geography of complexity*: what perhaps distinguishes it best from other perspectives is the fact that the observer is included in the observation – and that it recognizes that visible maps cannot be traced unless the invisible maps which write the modalities of description are first explored (Olsson 1991).

It remains to be seen, however, whether such a geography is feasible at the epistemological and practical levels. From an epistemological point of view, it is based within what Bocchi and Cerruti (1985) have termed the complexity paradigm. In particular, the adoption of both global-external and local-internal points of view implies the possibility of representing both the networks and their individual (local) nodes as systems open to any type of exchange – though ones which are operationally closed, i.e. essentially self-referential as far as the rules of their internal workings are concerned, and thus self-governing, aimed essentially at reproducing their own identity (Maturana and Varela 1980; Dumochel and Dupuy 1983). As a consequence, local nodes are not simply 'parts' of the network (i.e. subsystems whose properties are reducible to those of the global system) – and the global network is not simply the sum of the local parts of which it is composed. Not only is the local unable to exist without the global, but the global, too, is also dependent upon the very functioning of the various local systems.

There remains, of course, the much more difficult practical problem of substituting hypermodern geographies with complex ones. The success of hypermodern geographies is certainly linked to the performative force of their representations. The emphasis placed upon global leveling conditions, upon the speed and inevitability of change, the

fragmentation of places and actors, upon the absence of local constraints and regulations, often translates into a *de facto* legitimation of
today's dominant social and political order and, at the same time, into
a program for an ever more capillary colonization of the planet, aimed
at eliminating all forms of local resistance. In this fashion, any and all
processes of contestation can be reduced to but one (global) component,
with little probability of realization. However, there appears to be no
existing project capable of rebalancing the lack of symmetry in social
and cultural relationships between the global and local levels.

We can turn, for example, to the longstanding problem of territorial
justice. In a geography that recognizes the complexity of the relationship between global networks and local nodes, local underdevelopment
is seen, above all, as a defect of what Maturana and Varela (1980)
call the 'structural coupling' of the two systems. This 'defect' makes
it impossible to transform local values into values exchangeable within
global networks; not because local values do not exist (as would appear
when observing the system from the outside), but because the relationships framed by existing networks prevent local values from becoming
the values of global exchange. The solution in this case is: "If a man is
hungry, don't feed him fish, nor teach him how to fish, but exchange
your fish for something else he can produce." Or rather: do not try to
reduce the local to but one part of the (global) system, but allow it to
participate in an autonomous fashion. This signifies considering that
which is specific to the Other as something that we lack – and can benefit
from. From this point of view, territorial equity becomes not simply
equality or an elimination of difference but, rather, a *valorization of
diversity*.

All this, of course, relies upon the possibility of reciprocal translation
(or transcodification) of local languages, on their practical comprehension, and upon the possibility of sharing values produced in different
contexts and within unique experiences; a possibility which presents a
series of complex theoretical problems long debated also in the field of
geographical studies.[4] Theoretical quandaries which raise no less serious
practical considerations. Why, in today's world, has it become not only
difficult to appreciate (and therefore to desire) difference – but even to
tolerate it? To consider this a problem that can be resolved at a purely
pedagogical level would be to underestimate the disymmetry of the
'power geometries' (Massey 1993) which prevent most local systems
from taking an active part in the construction of (global) communication and exchange networks. At the global scale, in fact, representational spaces appear to be saturated by a single language and a dominant
value system, which not only represents itself but also claims to
represent all local realities as parts of its own 'self' in a relationship of

necessity/contingency. This sort of performative representation prevents local diversity from being recognized as such and thus reduces local systems' requests to but the right to self-representation in their own dialect. Such a representation, however, precludes the possibility of transcodification and, therefore, the possibility of participating within a global network of exchanges all the while conserving one's own local identity; of having such an identity fully recognized. And, as none of the solutions proposed in the field of geographical-territorial studies thus far appear totally convincing, the problem remains essentially open: presenting itself as the crucial question for a postmodern geographical praxis.

NOTES

1 'Specific' is intended here as that which is 'typical' of a given local urban system – that which confers it an advantage precisely because it is absent elsewhere or present in a very limited number of cases (for example, certain outstanding natural or historical/cultural conditions for tourism or the local accumulation of specific know-how in *milieux innovateurs*, etc.) Obviously, globalization also exploits *non*specific local resources (labor, primary resources, etc.) but in these cases the relationship with the global is one of dependency, local self-organization is not necessary, and the tendency is towards standardization (at least in the initial phases). Other authors, like Harvey (1993), though acknowledging the existence of processes of place construction assert, however, that globalization tends towards "cultural homogenization through diversification" in all cases.

2 This is certainly not the proper place to dissect in depth the epistemological aspects of the question, which I have examined elsewhere (Dematteis 1985). Suffice it to say that the approach can be traced back to the general scheme of abduction (Eco 1980b: 112ff.) and that it finds close parallels with the analogical use of the metaphor in the 'normal sciences,' with particular reference to the interpretations of M. B. Hesse (1970) and R. Boyd (1979). According to these authors, if we take proposition A' on system A (of, for example, social relations), then it is implicit in the metaphorical proposition B' (relating, for example, to B forms of 'geographical space') that between A and B there has to be some relation of analogy, similarity, or opposition – though not a relation of determination (such that A' can be deduced from B'), since if this were true, the metaphor would be immediately explicit and thus scientifically weak: that is, its analytical examination would not yield any new understanding. Rather, metaphor B' must imply a proposition (or set of propositions) M, which has relations of approximate equivalence (Hesse) or of partial denotation (Boyd) with A', comprehensible in intersubjective communication. It is in this zone of ambiguity that the metaphor functions as a model of analogical inference, capable of transfer-

ring to A ideas and implications normally associated with B (and vice versa), starting from a relation of analogy between A and B whose extension is originally only guessed but neither predictable nor describable with precision. According to R. Boyd (1979), in fact, the history of science presents numerous cases in which partially denoted terms have facilitated the discovery of new and significant features of the world.

3   The following refers back to my earlier essay "Global and Local Geographies" (Dematteis 1994)

4   See, among others, Mondada and Söderström (1991); Pred (1990).

REFERENCES

Agnew, J. A. (1987). *Place and Politics: The Geographical Mediation of State and Society.* London: Allen & Unwin.

Bagnasco, A. and P. Le Galès, eds. (1997). *Villes en Europe.* Paris: La Découverte.

Berry, B. J. L. and H. M. Kim (1993). Challenges to the Monocentric Model. *Geographical Analysis* 25: 1–4.

Bocchi G. and G. Cerruti, eds. (1985). *La sfida della complessità.* Milan: Feltrinelli.

Boeri, S., A. Lanzani, and E. Marini (1993). *Il territorio che cambia. Ambienti, paesaggi e immagini della regione milanese.* Milan: Abitare Segesta.

Boyd, R. (1979). Metaphor and Theory Change: What is 'Metaphor' a Metaphor For? In *Metaphor and Thought*, ed. A. Ortony. Cambridge: Cambridge University Press.

Cattan, N., D. Pumain, C. Rozenblat, and T. Saint-Julien (1994). *Le système des villes européennes.* Paris: Anthropos.

Conti, S., G. Dematteis, and C. Emanuel (1995). The Development of Areal and Network Systems. In *Urban Networks*, eds. G. Dematteis and V. Guarrasi. Bologna: Pátron, 45–68.

Dear, M. (1986). Postmodernism and Planning. In *Environment and Planning D: Society and Space* 4: 367–84.

Dematteis, G. (1985). *Le metafore della Terra.* Milan: Feltrinelli.

Dematteis, G. (1994). Global and Local Geo-graphies. In *Limits of Representation*, eds. F. Farinelli, G. Olsson, and D. Reichert, Munich: Accedo, 199–214.

Dematteis, G. (1997). Représentation spatiales de l'urbanisation européenne. In *Villes en Europe*, eds. A. Bagnasco and P. Le Galès. Paris: La Découverte, 67–96.

Dumochel, P. and J.-P. Dupuy (1983). *L'auto-organisation: de la physique au politique.* Paris: Seuil.

Eco, V. (1980). Metafora. In *Enciclopedia.* Turin: Einaudi.

Friedmann, J. (1978). The Urban Field as Human Habitat. In *Systems of Cities: Readings on Structure and Growth Policy*, eds. L. S. Bourne and J. W. Simmons. New York: Oxford University Press, 42–52.

Friedmann, J. and J. Miller (1965). The Urban Field. *Journal of the American Institute of Planners* 31: 312–20.

Garreau, J. (1991). *Edge City: Life on the New Frontier.* New York: Doubleday.

Governa, F. (1997). *Il milieu urbano: l'identità territoriale nei processi di sviluppo.* Milan: F. Angeli.

Hannerz, U. (1996). *Transnational Connections.* London: Routledge.

Harvey, D. (1993). From Space to Place and Back Again: Reflection on the Condition of Postmodernity. In *Mapping the Futures: Local Cultures, Global Change*, eds. J. Bird et al. London: Routledge, 3–29.

Hesse, M. B. (1970). *Models and Analogies in Science.* Notre Dame: University of Notre Dame Press.

Indovina, F. (1990). *La città diffusa.* Venice: DAEST.

Massey, D. (1993). Power-Geometry and a Progressive Sense of Place. In *Mapping the Futures: Local Cultures, Global Change*, eds. J. Bird et al. London: Routledge, 59–69.

Maturana, H. R. and F. J. Varela (1980). *Autopoiesis and Cognition: The Realization of the Living.* Dordrecht: D. Reidel.

Minca, C. and M. Dear (1997). Relativismo postmoderno e prassi geografica. *Rivista Geografica Italiana* 104: 277–303.

Mondada, L. and O. Söderström. (1991). Communication et espace: perspectives théoriques et enjeux sociaux. *Cahiers du DLS Lausanne* 11: 107–61.

Olsson, G. (1991). *Lines of Power: Limits of Language.* Minneapolis: University of Minnesota Press.

Pred, A. (1990). In Other Worlds: Fragmented and Integrated Observations on Gendered Language, Gendered Spaces and Local Transformation. *Antipode* 22: 33–52.

Racine, J. B. (1967). Exurbanisation et métamorphisme périurbain: introduction à l'étude du Grand-Montreal. *Revue Géographique de Montreal* 12: 313–41.

Raffestin, C. (1981). *Pour une géographie du pouvoir.* Paris: Litec.

Sassen, S. (1998). The City: Strategic Site, New Frontier. In INURA, *Possible Urban Worlds: Urban Strategies at the End of the 20th Century.* Basel: Birkhäuser, 192–9.

Sack, R. D. (1986). *Human Territoriality.* Cambridge: Cambridge University Press.

Soja, E. W. (1989). *Postmodern Geographies: The Reassertion of Space in Critical Social Theory.* London: Verso.

Turco, A. (1988). *Verso una teoria geografica della complessità.* Milan: Unicopli.

Weltz, P. (1996). *Mondialisation: Villes et territoires: l'économie d'archipel.* Paris: Presses Universitaires Françaises.

6

Adventures of a Barong: A Worm's-eye View of Global Formation

Steven Flusty

Globalization is a neologism that has become impossible to avoid. One finds it spilling from the lips of World Bank executives and Chiapan revolutionaries alike. The term's pervasiveness has rendered whatever it may refer to a likely causative suspect in occurrences that range from worldwide financial crises to neighborhood demographic change. But what *is* globalization?

Numerous theoretical formulations of globalization have emerged over the past decades. Some focus on the economics of international divisions of labor, others on worldwide sociological processes of cultural relativization, and yet others on representations of the world as a palimpsest. All these models of the global, whether foregrounding material, cultural, or discursive processes, demonstrate a privileging of structural levels of analysis. Even when academics undertake efforts to 'bring globalization down to earth,' such efforts commonly entail looking at gross aggregate data on migratory demographics, capital circulation, ethnonational composition, and the like. This perspective is predisposed to conceptualizing globalization as a system of higher-order processes. These processes are then deployed to account for the increasingly transnationalized and intercultural experiences, responses, and resistances of the quotidian. Viewed from such an elevated position, everyday occurrences become merely colorful anecdotes. The places in which the everyday occurs all too often become abstract surfaces, substrates that undergo (and resist) subordination to new transport and communication technologies.

Approaching globalization as a distinct object of study, a thing-in-itself that generates repercussions, is not inherently mistaken or without applicability. This perspective, however, conceals the extent to which globalization is em-placed, and the extent to which *em-placed* individuals and institutions are the capillaries of globalization at the same time as they are conditioned by it. Thus, globalization takes on the aspect of a juggernaut, traveling at warp-speed with nobody identifiable at the helm, leaving only the options of assimilation or resistance. I will argue that far from being an autocephalic process relentlessly bearing down upon us, globalization is produced, continually reproduced, and frequently redirected by means of everyday activities. In the process, I will demonstrate that conceptions of globalization as a graspable (and even wieldable) object are not enough. They must be augmented by an understanding of globalization as increasingly immanent in our commonplace thoughts and actions. Further, such an understanding renders globalization amenable to reforging through those same thoughts and actions.

My argument is divided into two parts. First, I review the preeminent and most geographically informed models of globalization, and their subsequent extensions into more locally grounded terrain. Second, I derive an alternative approach to globalization by deploying two artefacts and the social networks they inform. This alternative approach proceeds from the assumption that globalizing processes are human endeavors and, thus, may be fruitfully analyzed from the starting-point of the personal and the particular.

Globalization Defined?

The first use of the words 'globalize' and 'globalization' dates to the mid-1950s, where the terms emerged in a business context characterized by the growth of international trade (*Oxford English Dictionary*, 1988). Central to this growth was the post-Second World War rise of transborder investment that culminated in the emergence of readily relocatable 'quicksilver capital' and the rise of the multinational corporation (Pitelis and Sugden 1991). In the context of these developments, globalization has come to refer to the internationally coordinated command and control of the production and dissemination of commodities, capital, and information. There is ongoing debate over the desirability of this process. Some see it as the harbinger of a new commodity capitalism that facilitates an allegedly generalizable high material standard of living. Others decry it as a process of sociocultural standardization that benefits the already privileged while corroding local cultural variation informing alternative ways of life (e.g. Barber 1996; Barnet and Cavanagh 1994; Eisner and Lang 1991; Mattelart 1983; see also Dunning 1993 for claims of multinational corporations' new-found sensitivities to cultural diversity). Globalization itself, however, is taken as a given in such accounts, an inevitability to be either celebrated or lamented.

Others question this naturalizing of globalization. From this standpoint, international economic integration is predicated upon continuing terms of unequal exchange that favor the former colonial powers at the expense of the ex-colonial world. Thus, globalization becomes a discourse for legitimizing neocolonial practices that render ex-colonial societies penetrable to large financial and international aid institutions (McMichael 1996; Marcos 1995; see also Krugman and Venables 1995).

Immanuel Wallerstein has advanced the world systems model of globalization to expand upon this critique. Wallerstein argues that globalization is the perpetuation of colonial-style forced dependency through the global dissemination of a capitalist division of labor. The internationalization of this division of labor opens new markets and new proletariats for the expanded extraction of surplus value. Thus, labor exploitation is widely exported, enriching established classes of owners. Simultaneously, such a geographically expanded capitalism forestalls cyclical crises, spatially remediating the contradictions supposedly inherent in capitalist accumulation itself (Wallerstein 1984).

Corollary to this process is global alienation. As workers are separated a world away from the product of their work, what results is a worldwide commodity fetishism (the perception that products exist independently of their production processes). This gives rise to a global commodity culture ripe for penetration by such institutions as MTV, where meanings are assigned manipulatively to objects for marketing purposes. Thus, the internationalized mode of material-economic production generates the cultural aspects of globalization as an epiphenomenon (Wallerstein 1990).

The immediate end-products of this historical process are planet-length, readily relocatable commodity chains, linked place-nodes in which a given product's production occurs (Gereffi and Korzeniewicz 1994). These elongated commodity chains engender the quasi-ageographical flexible accumulation of disorganized capitalism (Lash and Urry 1987). The long-term end-product is, arguably, the global spread of class conflict and the planet-wide synchronization of capitalist crises, leading to the rise of world socialism.

Globalization defined by world systems theory is thus a process of continuing incorporation into a single economic regime. This incorporation is accomplished by means of an explicit spatial strategy wherein extensive commodity chains anchor the division of labor in each geographical periphery. Thus, mode of production yokes peripheries to core regions either directly or via the mediation of semiperipheral locales (Weisband 1989).

Roland Robertson's translocal theory of globalization adopts a more sociological stance. Robertson examines how societies and their members construct one another, and how this construction process is influenced by contact with other, equally constructed, societies.

The latter question of how societies construct themselves relative to one another becomes central to globalization in translocal theory due to 'compression.' Compression refers to the growing interdependencies between peoples and places. These interdependencies structure the world as a whole and bring societies into increasingly interpenetrative contact. Through compression, previously separated societies are brought (or forced) into contact so as to heighten perceptions of the differences and similarities between each. Thus, translocal globalization may be seen as predicated upon conceptions of the spatial separation of more-or-less distinct culture areas, and of the widespread local impacts of bridging these separations with increasing frequency and pervasiveness.

This dynamic intensifies the potentials within the global field. For Robertson, the global field consists of four elements: the self, the national society, the world system of societies, and humankind. With compression, he claims, the interrelationships of these elements attain a new prominence, thus heightening individuals' and societies' consciousness of humankind and a world system of societies. This new consciousness is thus one of relativization. It forces the continual redefinition of citizenship, one's cultural identity, the unique value of a nation in comparison to others, and the very existence of national society (Robertson 1992). Such redefinition spawns diverse local outcomes like internationalized 'Post-Fordist people' (Barns 1991), 'Occidentalism' born of Japan's sense of technological accomplishment (Morley and Robins 1992), or the fall of whole nation-states too atomized or insular to adapt to a global context (Tuan 1996).

Thus, globalization as defined by translocal theory is a process of conceptual exchange between the boundaries that separate numerous societal regimes (or the refusal of such exchange; see Perrin 1979). These exchanges define each society and its members relative to one another, and so establish worldwide a field of negotiated difference. Within this field, the identification of individual and social uniqueness becomes the universal norm. Simultaneously, universality, the search for global fundamentals, becomes a particularistic concern embodied in a plethora of movements advancing their own 'real meanings' of the world (Robertson 1992: 177–8).

Arjun Appadurai's (1990) chaotic theory of globalization asserts that both the economic and the cultural realm, along with others, are produced by more fundamental turbulent flows of matter and mentation.

These flows are a product of logistical improvements and related technologies overcoming time and distance. With these impediments to communication and the command of resources removed, cultural transactions increase between social groups that have been previously separated either by circumstance (e.g. geographical barriers) or intent.

The consolidation of planet-wide contact has resulted in the sustained circulation of commodities, data, and ideas, most importantly ideas of peoplehood and selfdom. According to Appadurai, these flows create five dimensions of flow experienced by actors as diverse as individuals, families, neighborhoods, diasporic communities, and even nations and multinational institutions. Appadurai refers to these dimensions as landscapes, or 'scapes,' in recognition of the fact that they are perspectival constructs with fluid and irregular shapes. The five landscapes are: ethnoscapes, the interconnections of social affinity groups; mediascapes, the distribution of the capacity to disseminate information and mediated images of the world; technoscapes, the global configuration and flow of technologies; finanscapes, the shifting global disposition of capital; and ideoscapes, the global dispersion of political-ideological constructs, generally statist or counter-state and couched in Enlightenment notions of 'freedom,' 'welfare,' 'democracy,' etc.

Because these flows are present with differing degrees of intensity at different locations, experiences and perceptions of them differ radically. Such differentiated experiences of these "scapes" will lead to people inhabiting equally differentiated 'imagined worlds,' parallel visions of the world potent enough to subvert the orderings of 'the official mind' (Appadurai 1990: 7). It is the propagation of these imagined worlds and the differences between them, overlain with the disjunctures between the global 'scapes,' that constitutes the process of contemporary globalization.

Globalization in Appadurai's chaos theory is thus distinct types of highly irregular material and immaterial flows. These flows coagulate differentially in specific places, generating a multiplicity of disjunctive domains that are fractal in shape and "polythetically overlapping in their coverage of terrestrial space . . ." (Appadurai 1990: 20).

Assessing Global Representations

Each of these theories provides unique and valuable insights into processes of globalization. Further, while these takes on the 'object' that is globalization are so divergent as to be largely incommensurable, this affords the possibility of applying them simultaneously to produce polyperspectival analyses of 'the global.' Each theory also has its specific weaknesses and exclusions. World-systems theory leaves little room for the persistence and renewed vigor of particularistic sociocultural resistance rooted in local identity. Translocal theory evinces greater cognizance of such cultural dimensions, but is also prone to reproducing some of the cultural area hypothesis' most notorious essentializing tendencies. A particularly egregious example of such essentialisation is Samuel Huntington's adaptation of relativization to geopolitics. His scenario hypothesizes the existence of distinct Western, Confucian, Japanese, Islamic, Hindu, Slavic-Orthodox, Latin American, "and possibly African" civilizations, and foresees them entering into economic and even armed conflict against one another (Huntington 1993; see also Ajami's response 1993). For its part, chaos theory accommodates both economic and cultural dimensions within its multiple semi- and discontiguous global circuits. It also fails to map these circuits or apply them systematically, leaving them as suggestive topological metaphors.

A broader criticism against these perspectives can be targeted at their shared conceptualization of globalization as a 'thing.' Each approach assumes that globalization is *out there*, a result of the inevitable dynamics of capital, technology, and the planet imploding under its own social weight. It is something that happens to us, and is largely beyond our control. This assumption leaves us to contend with globalization, to adapt or resist, as best we can with a limited range of options. The commonplace is considered symptomatic of globalization, rather than as productive of globalization. Simultaneously, space is confined to the role of something upon (and against) which globalization acts. As a result, and with the partial exception of Appadurai's approach, all these perspectives envision a process of globalization in which daily life and its places are a residual product.

This imbalance must be redressed by conceptions of globalization that look to specific in situ everyday activities. Further, such activities must be treated not just as informed by the global, but as formative of the global. Cribbing from Foucault's critiques of the nature of power, focus must shift away from an extrinsic, 'sovereign' globalization towards a 'nonsovereign' globalization. Such a 'nonsovereign' globalization is as embedded in, and emergent through, our individual day-to-day machinations as it is in the mass movement of capital, populations, and information.

There have been some tantalizing glimpses into what 'nonsovereign' globalization may entail. Massey (in Bird 1993), for example, argues that many individual spaces are unique articulations of particular global influences, and that such spaces have multiple identities "that can be either, or both, a source of richness or a source of conflict." She finds such a place in Kilburn High Road, on the outskirts of London. There, posters in support of the Irish Republican Army coexist with Indian sari shops, a Muslim merchant depressed over the Persian Gulf War, and traffic congestion generated by increased business travel at Heathrow airport.

Sassen is similarly attuned to the eccentric geographies of the emergent global. Diversifying common definitions of the 'world city,' she points out the existence of places that are:

> sites not only for global capital, but also for the transnationalisation of labour and the formation of transnational identities . . . the terrain where people from many different countries are most likely to meet and a multiplicity of cultures come together. (Sassen 1996: 217)

Such places are evinced by the contrasts between the simultaneous presence of "the corporate city of high-rise office buildings, the old dying industrial city, the immigrant city," experienced as "[a] space of power; a space of labor and machines; a Third World space" (Sassen in King 1996: 23). The inhabitants of these spaces, and their quotidian undertakings, are no less polyvalent. According to Sassen, one finds elites who "think of themselves as cosmopolitan" and members of "localized" cultures who, in coming "from places with great cultural diversity," may themselves "be as cosmopolitan as elites." Many of these inhabitants find themselves with "unmoored identities" and, while undergoing reterritorialization in new settings, invent "new notions of community, of membership . . . of entitlement," and a new transnational politics to match (Sassen 1996: 217–19).

Massey and Sassen go a long way towards pointing out the lived specificities and variegated terrains of globalization. They present these, however, as local manifestations of globalization that, while sufficiently potent to inflect global formation, remain the byproduct of larger transnational flows of investment capital and immigration. But what if the minutiae of global formation were approached as the underpinnings of globalization, rather than as the result? What if global formation were reimagined as the result of Massey's unique articulations and Sassen's places of contrast filled with complex and unmoored identities? In short, what remains lacking is a serious investigation and ground-up theorization of the specific, commonplace, lived geographies of globalization. The remainder of this paper will take an initial step towards just such an investigation.

Working with 'Stuff'

How might the everyday processes of global formation be studied? In a photographic essay and accompanying text, Allan Sekula has suggested a particularly inspired approach. In *Fish Story*, Sekula assumes that globalization is most clearly legible in the transnational passage of concrete artefacts between people who send and receive them, and in the material practices of transporting these artefacts. Working from this assumption, Sekula places himself aboard a containerized cargo ship, and documents the daily lives of the ship's crew, dockworkers at various ports, and ways of life adjacent to these ports. Thus, he employs inter-modal cargo containers to trace networks that incorporate such diverse elements as Taiwanese and Filipino sailors, Eastern European bridge officers, robotic cargo handlers on the German coast of the North Sea, fishermen living alongside Mexican ports in cast-off cargo containers, and a host of others. *Fish Story* thus reveals artefacts, and the relation-ships accounting for and influenced by their production and dissemina-tion, as both a linchpin in globalization and a powerful tool for the study of globalization.

Despite this significance, artefacts constitute a strangely invisible presence in our lives. We interact with thousands of artefacts every day from the moment we take our head off the *pillow*, sit up in bed, and put our feet on the *rug*. Yet the overwhelming majority of such artefacts are so commonplace, so deeply embedded in our surroundings and routine activities, that we seldom notice more than a fraction of them.

This relative anonymity, however, hides power and complexity. Arte-facts are a unique form of material evidence, possessed of a twofold nature (Miller 1987: ch. 7). On the one hand, a given artefact exhibits a recognizable form regardless of place or mode of employ. On the other hand, the same artefact is attached to differing functions and meanings that depend upon the social context of its use. Thus, the artefact is simul-taneously thing and image. But it is more than a thing to be utilized and filled with meaning. As per Sauer and Boas (in Agnew et al. 1996), it is the product of unique human/environment relationships, and it is an indicator of the geographical extents of particular cultural influences.

It is also an active mediator of people's relations to one another and their environment (Schlereth 1990; see also Kingery 1998: ch. 1). Material culture is not the passive embodiment of human intentions, but "a cause, medium and a consequence of social relationships" (Riggins 1994: 1). It is a synecdoche for a larger society, an active component of social reproduction that exhibits its own agency and even the physical environment within, and upon which, societies sustain themselves (Jackson 1984).

Thus, artefacts may be read as mediators and participants in every-day global formation, provided that the artefact's material and social faces are made visible. Material and narrative culture must be consid-ered jointly. It is not enough to track the passage of artefacts. Mappings of transnational artefact flows must be joined to the stories behind the generation, transportation, and use of those artefacts. Central to this joining is a redefinition of the artefact itself. The artefact must cease to be taken as a crystallization of absent social processes. Instead, it must be read as a constituent component of its material and social contexts:

> we have to follow the things themselves, for their meanings are inscribed in their forms, their uses, their trajectories. It is only through the analysis of these trajectories that we can interpret the human transactions and calculations that enliven things. (Appadurai 1986)

By redefining objects not just as denoting relationships, but also as participants in relationships, we acknowledge a narrative material culture that erodes firm divisions between user and thing used. This enables us to simultaneously generate social histories and cultural biographies of people and things together (Kopytoff in Appadurai 1986). Such narrative material culture reveals artefacts embedded in diverse and disparate (sub)cultural narratives and locales. It also shows how those narratives are linked by the increasingly translocalized activities of daily life and, thus, are not so disparate at all.

In following the artefact as it traces circuits linking lives lived at particular addresses and in varied contexts, this approach dovetails with Actor-Network Theory (ANT). ANT is predicated upon the supposition that humans and nonhumans enter into networks of relationships, and that such relationships change each component member through processes of mutual translation. The resulting networks, hybridized of co-adapted humans and nonhumans, may be considered complex agents in their own right (Latour 1988). ANT's network entities are thus very similar to the human/artefact hybrid objects constituted in a narrative material culture. ANT, however, is explicitly ageographical, asserting that there is nothing and no space outside the lines and points of the network (Latour 1997). Narrative material culture, by way of contrast, focuses on the artefact's locales of appearance, and on its influences in physical space and individually experienced place. This serves to emphasize that human/nonhuman interactions comprise "social practices [that] are inherently spatial, at every scale and all sites of human behavior" (Dear and Wolch 1989: 9).

I will now employ material culture to construct two narrative mappings of complex and extensive commonplace relationships. Such relationships constitute hybrid global networks of particular people, places, activities, and artefacts. Thus, the narrative mappings that follow will describe globalized social geographies of people and things together. In the process, they collapse 'the global' into everyday life.

The Footloose Barong

The barong tagalog, or barong for short, is a Filipino men's dress tunic with collar, cuffs, and a side vent at either hip. It is similar to the Caribbean-cum-Yucateccan guayabera 'wedding' shirt, and to other formal-dress garments diffused across large portions of the globe by the Spanish empire. The barong is traditionally made of 'jussi' pineapple fibre, the pineapple having been introduced from South America almost five centuries ago by that same Spanish empire. Organza is now often substituted for jussi, both being light, transparent fabrics well suited to the heat and humidity of the Philippines. Barongs are worn on special occasions by street vendors and the President alike, untucked and long, usually over a crew- or V-necked white cotton undershirt.

The significance of the barong here, however, is less a matter of what it is, than of how I ended up getting one. I had wanted a barong ever since I had first seen one during a brief trip to Hawaii in the 1970s, but only succeeded in obtaining a barong in 1997. And in a bow to my own place of origin, Los Angeles, this particular barong's travels and ultimate acquisition were inadvertently initiated by my automobile.

My car is a 1987 Nissan pick-up, a 'Japanese' truck assembled in Smyrna, Tennessee, of parts originating in more than half a dozen countries. Ever since I purchased the truck, I have taken it for service at Imperial Automotive, largely because Imperial is only a few blocks from my home. For as long as Imperial has existed, it has been owned by Rob, a member of the Armenian diaspora. His family was deported to Syria in 1915, when the Armenians on the Ottoman empire's northern frontier were judged detrimental to the 'Young Turks' nation-building project. Seen as outsiders in Syria and thus denied higher education, Rob's family pursued a trade and (like many Armenian refugees in the Middle East) became automobile mechanics. Many of his family members carried that trade with them to the United States, ultimately comprising the present ownership and staff of Imperial.

During one visit for a tune-up and oil change, I expressed interest in the dance music being played in the garage. Rob filled me in on the performer, a Los Angeles based Irani-Armenian known as Avak. Avak is something of the 'Michael Jackson of Persian-pop,' famous for performing such Persian popular music hits as Gloria Gaynor's "I Will Survive." Such music having been banned from Iran's Islamic Republic, Avak and his band shuttle between particular neighborhoods of Los Angeles, Manhattan, London, Milan, Frankfurt, Paris, Stockholm, and Dubai. With the collapse of the Pahlavi Dynasty in the late 1970s, they have became home to an affluent, urbanized class of Irani professionals. In the process, these cities have formed a circuit in which Avak and similar musicians stage concerts.

This concert circuit includes the Assyrian Hall some eight miles from my house, where I attended one of Avak's performances at Rob's recommendation. There, I met Avak's bassist: Sheryl, a second-generation Japanese-American, hired for her musical talents and retained for the rapidity with which she learned to rap in Farsi. We became close friends. On one occasion, as we sat idly channel-surfing, we paused on an international station featuring a newscast from the Philippines, anchored by a newsreader dressed in a barong. I commented upon how I had long wanted such a shirt, and we proceeded on to other topics of conversation, and to other channels.

At this juncture of my barong's journey, its continuing adventures hinge upon a more complete description of Sheryl herself. In addition to all her previously-mentioned attributes, Sheryl is, in her own words, 'a dyke.' One result of this has been Sheryl's predilection for a coffeehouse called Little Frida's, named for the Mexican-Hungarian-Jewish artist Frida Kahlo. Little Frida's is some 20 miles from my house in the City of West Hollywood, an independent city embedded within Los Angeles. West Hollywood is noteworthy as a mecca for gay men, but includes a small collection of women's bars and lesbian-owned businesses, of which Frida's is one of the most prominent. It was at Frida's that Sheryl befriended Moire, a self-described "Filipina lesbian feminist." Moire was also an employee of AT&T telecommunications corporation, selling discounted phone access accounts to accommodate those Asians in the United States and Asia in need of sustained transpacific contact. As a favor to Sheryl, Moire used her AT&T account to contact a sibling visiting family in Manila, roughly 7,000 miles from my house. This sibling returned with the barong and handed it over to Moire. Moire passed the barong along to Sheryl. And Sheryl gave the shirt to me.

Now, in this expanding network of day-to-day relations I have delineated, the barong is just one of a plethora of possible protagonists. We could branch off with Avak's music, and discuss how his sense of 'Armenian-ness' has led him to approve the circulation of pirated cassette tapes amongst cash-strapped fans in Yerevan. Or we could follow an edgy subset of Little Frida's habitués a few blocks east to the Pleasure Chest, an 'adult novelties' emporium that has responded to neighborhood demographic change by displaying 'Parking for Customers Only' signs in Russian.

All of these parallel stories reinforce the quotidian processes of global formation exemplified by my barong's adventures, processes real enough to put the very shirt on my back. This shirt came to me not *just* on account of an internationalized division of labor, or a relativizing inter-cultural flow of symbolic forms, or a collision of discursive realms prob-lematizing construction of 'the Other.' I acquired the shirt on account of specific people who, in the course of negotiating their daily lives, hap-pened to find themselves coming together in particular places.

None of this is to suggest that these em-placed people are not materially, symbolically, and discursively constituted. But central to my story are these people's daily lives, and how those lives are spatially sit-uated. My conspirators in this tale are each members of collective social formations, whether technical affiliations like musicians, ethnic affilia-tions like Armenians, or marginalized affiliations like lesbians. These various social collectives are both globally dispersed and technologically interconnected. Simultaneously, they are clustered together in the space of a single world city. Thus they inform intra-urban overlappings of inter-urban, international communities, and create worldwide channels for the dissemination of symbols and material goods.

Enmeshed in these channels are intimate experiential dimensions that highlight the much overtaxed term *difference*. The story I have sketched consists of differences *that occur* due to their relational coming together in particular places. In this coming together we witness the prolifera-tion of a special kind of concrete place: a place where differences must be actively translated into some mutually approximable language if commonplace social interactions are to occur. Such places of translation materialize at multiple sites as public as a garage where the owner of a 'Japanese' vehicle is exposed to Persian pop music, or a coffee-house where the same vehicle-owner becomes a single social unit with a bassist and a Filipina.

Places of translation materialize at multiple scales as well. These include places as intimately private as the person of a Filipina lesbian feminist employed by a multinational telecommunications corporation. This points to the fact that the everyday 'comings together' that gener-ate the global occur at both interpersonal and intrapersonal levels. It is significant that the conspirators in acquiring my barong are members of propinquitous, globalized social collectivities. But it is equally signifi-cant that each of these conspirators is simultaneously a member of more than one such collective. Each individual in my story possesses multi-ple identities. Thus, each acts as a point of contact between worlds. Each is a bodily site where pluralities of worlds overlap to facilitate an inter-penetration of social formations. Such embodied points of contact demonstrate that everyday processes of global formation have not just an interpersonal dimension, but an intrapersonal one as well. In the absence of this intrapersonal dimension, here would be no connections between the numerous social terrains separating me from the Philippines, and my barong's journey would have been impossible from the start. What, then, are the intrapersonal dynamics of em-placed, everyday globalization? To begin answering this question, I will invoke a second artefact.

Naming He who Did Not Come from Overseas

Tokunboh is a set of phonemes, arranged in a sequence unique to the Yoruba language. Together, these phonemes comprise a male proper name that translates roughly as "he who comes from overseas." While this name was once relatively unusual amongst the Yoruba people, it has become increasingly common throughout the twentieth century. Tokunboh's growing popularity is a product of this people's location within the Nigerian state. The colonial and postcolonial establishment of Nigeria placed the Yoruba within the same political territory as the Ibo, the Ishan, and more than 250 other peoples. Of these groups, the Yoruba exert a particularly strong cultural influence, exemplified by their founding of the country's primate city and former capital, Lagos. Further, solidarity amongst peoples like the commonly christianized Yoruba, Ibo, and Ishan has been intensified by their intermittent opposition, in both the British colonial and contemporary eras, to rule by the predominantly Muslim Hausa people. All these factors have resulted in intensifying cross-transfers of Yoruban culture, most notably language, music, and naming practices.

Events in Nigeria throughout the 1980s and 1990s have played an even larger role in popularizing the name Tokunboh. The end of the oil boom and the corollary introduction of International Monetary Fund Structural Adjustment Programs deeply disrupted the petroleum-dependent Nigerian economy. Faced with sudden prosperity that gave way to equally sudden economic collapse, there has emerged a sense that hopes for improved material living standards are futile, and that Nigeria's internal condition can only continue to deteriorate. This hopelessness has been exacerbated by a succession of military dictatorships notorious for rampant corruption and physical coercion. One Nigerian journalist, in self-imposed exile but nonetheless requesting anonymity for fear that family left behind will be "clapped into detention," is quite clear about this last point. Referring to the 1993 handover of power from the Babangida to Abacha regimes, she characterizes the succession as a change from a President "who'd smile while he'd kill you" to one "who won't even bother to smile." Economic and political realities in Nigeria have thus engendered massive waves of emigration. Many of these émigrés believe living conditions in Nigeria are so irreparably bad that there is no reason to ever return. Such immigrants comprise what will likely prove to be a permanent Nigerian diaspora. Thus, growing numbers of Nigerians have indeed come from overseas, rendering Tokunboh an especially popular name amongst diasporic Nigerians.

As with the barong, the significance of the name Tokunboh in this story is that, amongst certain social circles, it has become my name. I am not Nigerian, let alone Yoruba. And in relation to West Africa, I have yet to cross the sea. In fact, to receive the name I traveled no further than 100 feet, from my front door to the house across the street.

While London remains perhaps the locale of choice for Nigerian immigrants, many proceed on to the metropolitan centers of the United States. Of these US metropolitan centers, Los Angeles has gained particular prominence. As of the 1990 census, there were over 5,000 Nigerians resident in LA (Allen and Turner 1997: 65). According to guesstimates by members of Los Angeles' Nigerian communities, that number will more than quadruple in the 2000 census.

The Okonkwos are amongst these expatriates. Their multigenerational extended family is spread between Lagos and the West Coast of the US. Eighteen of them now reside in Los Angeles. In early 1997, a subset of these 18 moved into the house across the street from mine, and this house has become a site of regular visitation for the remainder. Similarly, it had become a site of regular visitation for me. On one of these visits, on Christmas Eve of 1997, I stumbled into a boisterous conversation conducted simultaneously in Yoruba, Ishan, Pidgin, and English. The topic of the conversation, insofar as I could follow it, was something about the appropriateness of one name versus another. Only when the conversation came to some sort of resolution, and I was presented with my Christmas gift, did I understand what had transpired. Because I was 'learning to be Nigerian,' as the Okonkwos saw it, I was given the name Tokunboh. The name caused particular amusement, because it inverted the Okonkwos' geographical mobility relative to my own.

Although the name was a gift, there was the tacit implication in its mode of presentation that it was also something I had earned, however inadvertently. The gift name was thus an acknowledgment of something I had not fully realized: over the course of my continual visits to the Okonkwo home, and subsequent outings around the city, I had gradually acculturated to what the Okonkwos considered intrinsically 'Nigerian' everyday practices. Gradually, I have learned that a party will start some one to four hours later than its announced time, and have adjusted my own sense of punctuality accordingly. I know that when the family matriarch calls and requests transportation to one of the local big-box retail outlets, it is incumbent upon me to accede. I ceased to see anything unusual in using my bare hands to eat stews with FuFu, a sticky pounded yam dough. I am familiar with the musical differences between Femi Kuti and his father, the late Fela Kuti, and I know how to dance to either. I can tell stories about the Shrine, a club founded in Lagos by Fela himself, despite never having visited the place. I find myself understanding a fair amount of conversational Pidgin. I have developed a visceral sense of the human rights abuses committed by multinational petrochemical companies prospecting for oil in Ogoniland. And I have participated in the boisterous cynicism when we gathered to watch the 1999 inaugural address of President Olusegun Obasanjo.

None of these things makes me either Yoruba or Nigerian, any more than donning a barong makes me a Filipino. Nor, in entering into dialogue with the Okonkwos, have I lost anything that I was prior to that dialogue, much as I lose nothing in encountering Avak's music. Rather, I have been augmented and, in some ways, reconfigured by new perspectives and new ways of being. In a sense, my ongoing dialogical engagement with the Okonkwos has simultaneously dis-placed and re-placed me, turning me into a point of contact. In the instance of this paper, for instance, I have become a place of translation between diasporic Nigerians and geographers.

Such diversification of individual identities renders them active translators that create conduits between social collectives. It is these diversified, translative identities that permit seemingly disparate social and geographical realms not merely to overlap, but to constructively inter-penetrate. 'The global' is thus formed in the everyday by practices that play along (and across) the physical and psychic boundaries of both widespread social collectives and their individual members. This contradicts the hypothesis that focusing on identity and difference must necessarily yield hostile divisiveness and a balkanized social terrain. Rather, the amity and invention inherent in the commonplace practices of global formation, at both intrapersonal and interpersonal levels, provide ample cause not just for investigation, but also for optimism.

Conclusion

Globalization is in flows of capital and waves of migration, in satellite broadcasts and in transoceanic air-routes. But the global is no less in the heads, and commonplace interactions, of those whose daily lives underpin these larger-order phenomena. Given the critical significance of globalization's interpersonal and intrapersonal dynamics, to characterize them as a 'worm's-eye view' of globalization may thus seem belittling. Indeed, I had contemplated revising this paper's subtitle to read 'an insider's view.' But as both my adventurous barong and my new name demonstrate, to do so would have been to commit a redundancy, as we are all insiders no matter how cosmopolitan or local we may appear. This is not to imply that we are all equal in the efficacy of our global reach. But can we say with certainty that the ultimate impact of a transnationally itinerant musician or a dissident émigré is any less real than that of a multinational corporation?

All views of the global are views from the inside. We are the capillaries of globalization, variously participating in complex webs of emerging relationships that are simultaneously spatially extensive and psychically intensive. We actively produce the 'local holes' through which the 'global flows' we hear so much about percolate or, more precisely, are percolated. Globalization is 'nonsovereign.' It is constituted by and it constitutes the human and nonhuman actors who stand on or sit in or move through concrete places while engaged in production and consumption, transporting and meaning-making.

Such actors seem to be puny things when contemplated from a distance sufficient to encompass multinational markets, world systems, global fields, and a riot of terrestrial scapes. By finding globalization in the circulation of a shirt or the granting of a name, however, it becomes apparent that human and nonhuman actors make the world. Without these actors, there can be no global flows to speak of, and thus no processes of global formation. And because of these actors, our comprehension of 'the global' must necessarily remain insufficient until we approach 'the global' from its most intimate basis in localized everyday existence.

NOTES

All names in this document have been changed, with the exception of the recently defunct Little Frida's. While 'Avak' is a public figure, I have modified his to maintain the (now ex-) bassist's privacy.

References

Agnew, J., D. A. Livingstone, and A. Rogers, eds. (1996). *Human Geography: An Essential Anthology*. Oxford: Blackwell.
Ajami, F. (1993). The Summoning. *Foreign Affairs* Sept./Oct.: 2–9.
Allen, J. P. and E. Turner (1997). *The Ethnic Quilt: Population Diversity in Southern California*. Northridge: The Center for Geographical Studies.
Appadurai, A., ed. (1986). *The Social Life of Things*. Cambridge: Cambridge University Press.

Appadurai, A. (1990). Disjuncture and Difference in the Global Cultural Economy. *Public Culture* 2(2), Spring: 1–24.

Barber, B. (1996). *Jihad vs. McWorld: How Globalism and Tribalism are Reshaping the World*. New York: Ballantine Books.

Barnet, R. J. and B. Cavanagh (1994). *Global Dreams: Imperial Corporations in the New World Order*. New York: Simon and Schuster.

Barns, I. (1991). Post Fordist People? Cultural Meaning of New Technoeconomic Systems. *Futures*, Nov.: 895–914.

Bird, J. et al., eds. (1993). *Mapping the Futures: Local Cultures, Global Change*. London: Routledge.

Dear, M. J. and J. R. Wolch, eds. (1989). *The Power of Geography: How Territory Shapes Social Life*. Boston: Unwin Hyman.

Dunning, J. H. (1993). *The Globalization of Business: the Challenge of the 1990s*. London: Routledge.

Eisner, M. and J. Lang (1991). It's a Small World After All/The Higher the Satellite, the Lower the Culture. *New Perspectives Quarterly* 8(4): 40–5.

Foucault, M. (1980). Two Lectures. In *Power/Knowledge: Selected Interviews and Other Writings, 1972–1977*, ed. C. Gordon. New York: Pantheon.

Gereffi, G. and M. Korzeniewicz, eds. (1994). *Commodity Chains and Global Capitalism*. Westport: Praeger.

Huntington, S. P. (1993). The Clash of Civilizations? *Foreign Affairs*, Summer: 22–49.

Jackson, J. B. (1984). *Discovering the Vernacular Landscape*. New Haven: Yale University Press.

King, A., ed. (1996). *Re-Presenting the City: Ethnicity, Capital and Culture in the 21st-Century Metropolis*. New York: New York University Press.

Kingery, W. D., ed. (1998). *Learning from Things: Method and Theory of Material Culture Studies*. Washington, DC: Smithsonian Institution Press.

Krugman, P. R. and A. J. Venables (1995). *Globalization and the Inequality of Nations*. Cambridge: National Bureau of Economic Research.

Lash, S. and J. Urry (1987). *The End of Organized Capitalism*. Oxford: Polity Press.

Latour, B. (as J. Johnson) (1988). Mixing Humans and Nonhumans Together: The Sociology of a Door Closer. *Social Problems* 35(3): 298–310.

Latour, B. (1997). On Actor-Network Theory: A Few Clarifications. Soziale Welt. http//www.keele.ac.uk/depts/stt/stt/ant/latour.html

McMichael, P. (1996). Globalization: Myths and Realities. *Rural Sociology* 61(1): 25–55.

Marcos, Subcomandante (1995). *Shadows of Tender Fury: The Letters and Communiqués of Subcommandante Marcos and the Zapatista Army of National Liberation*. New York: Monthly Review Press.

Mattelart, A. (1983). *Transnationals and the Third World: The Struggle for Culture*. South Hadley, Mass.: Bergin and Garvey.

Miller, D. (1987). *Material Culture and Mass Consumption*. Oxford: Blackwell.

Morley, D. and K. Robins (1992). Techno-Orientalism: Futures, Foreigners and Phobias. *New Formations*, Spring.

Perrin, N. (1979). *Giving Up the Gun: Japan's Reversion to the Sword, 1543–1879.* Boston: D. R. Godine.

Pitelis, C. N. and R. Sugden, eds. (1991). *The Nature of The Transnational Firm.* London: Routledge.

Riggins, S. H., ed. (1994). *The Socialness of Things: Essays on the Socio-Semiotics of Objects.* Berlin: Mouton de Gruyter & Co.

Robertson, R. (1992). *Globalization: Social Theory and Global Culture.* London: Sage.

Sassen, S. (1996). Whose City Is It? Globalization and the Formation of New Claims. *Public Culture* 8(2): 205–33.

Schlereth, T. J. (1990). *Cultural History and Material Culture: Everyday Life, Landscapes, Museums.* Ann Arbor: UMI Research Press.

Sekula, A. (1995). *Fish Story.* Dusseldorf: Richter Verlag.

Tuan, Y. (1996). *Cosmos and Hearth: A Cosmopolite's Viewpoint.* Minneapolis: University of Minnesota Press.

Wallerstein, I. (1984). *The Politics of the World-Economy.* Cambridge: Cambridge University Press.

Wallerstein, I. (1990). Societal Development, or Development of the World System? In *Globalization, Knowledge and Society.* London: Sage.

Weisband, E., ed. (1989). *Poverty Amidst Plenty: World Political Economy and Distributive Justice.* Boulder, Colo.: Westview.

7

Rescaling Politics: Geography, Globalism, and the New Urbanism

Neil Smith

I am beginning to wonder whether what I have always believed to be so centered is in fact dispersed. Have I become tired of the known? The more I encounter those who impart to me their knowledge of distance and time, the more I begin to believe the real object of my quest is to allow myself to become entranced.

Frà Mauro

Frà Mauro was a voracious scholar. He absorbed the knowledge of others as most mortals breathe air, and he wrote the world with such a careful and studied completeness that he eventually came to see himself as his own text. "At last I have composed the world," he concluded after a life's work. "I have sliced through its materiality. I have peeled back its corporeal skin to reveal its inner workings." But the map he thereby etched of the world's surface delivered none of the freedom he expected from its "unabashed spaciousness." "Gazing at the map, I begin to see a portrait of myself. All the diversity of the world is intimated" in the map, "even as this diversity is intimated within me. An aura of remoteness hovers about its contours . . . I have emptied matter of its content. Now it shimmers, diaphanous, a subtle body whose origins lie elsewhere."

Frà Mauro's dilemma is familiar. On the one hand it recalls Marx's analysis of the commodity and alienation: the product of his work takes on its own life. But more importantly, if he indulges a modern ambition to unpeel the world's inner workings, Mauro's revelation that he writes a diverse world and his own contradictions as a single script has a far more postmodern feel to it. The easy scalar interrelationality between global map and his *own* 'subtle body' only reaffirms the postmodern feel. In fact, if James Cowan is to be believed, Mauro was a sixteenth-century monk who closeted himself away in the Venetian monastery of

San Michele di Murano, and there received a constant stream of traders, travelers and emissaries who eagerly disgorged their knowledge of the places they had seen (Cowan 1996: 144–5; the epigraph is from p. 102). Mauro's entire meditation therefore lends powerful support to the notion that, historically and conceptually, the postmodern is ever already present in the modern. The postmodern refusal of origins is, in fact, a quite specific denial of postmodernism's own nativity in the origins of modernity.

From his perch on Venice's "lagoon of soupy canals, cats' pee and pageants," Mauro understood well that "mapmakers embroider the world." And his account of self-transformation during the decades-long effort to map the world mimics an intellectual journey from a rarefied proto-Cartesian idealism through heady contact with empirical knowledge to a skeptical postmodernism. He assumes the world as "a place entirely constructed from thought, ever changing" as the thinkers themselves change in response to new knowledge: "the world is not real save in the way each of us impresses upon it his own sensibility." But then he begins to doubt that his completed map is anything more than his own trace in the world; purely for self-entertainment he has indeed emptied the world of its content – "to allow myself to become entranced." Frustrated that this might be all there is to an intensely examined life, and that his completed map forever circumscribes his future movement, Mauro craves the opposite of his life work: "the kingdom of 'no-knowledge'" (Cowan 1996: 12, 21, 60, 132, 148). At root, a dualist modernism *is* the purest dilemma of postmodernism.

Throughout his attempt at global cartographic embroidery, Mauro insists that he wants to establish the outlines of global space but has no interest in the substance of that space. His impasse might stand as an analogy for much contemporary debate around modernism and postmodernism, but it may also apply to current debates over the local and the global. The localities debates that emerged in Anglo-American geography in the late 1980s involved the attempt to ally a geographical shift of scale, from global to local perspective, with an historical shift from modern to postmodern. Mauro embraces both positions. He assuredly constructs the global – self-consciously objective and alienated – but does so from the vantage point of his own radical, subjective isolation. (Not only does he always remain in Venezia; he inhabits a monastery.) Much the same might be said of the localism which emerged primarily in England in the 1980s, except that any influence it had was eclipsed by an emerging frenetic cacophony over 'globalization.' At the beginning of the twenty-first century, therefore, it is the norm to embody both positions, much as Mauro did five centuries earlier: the realities of globalization are matched by a powerful impetus toward localization that fuelled the locality debates in the first place. Erik Swyngedouw's ungainly language of 'glocalization' awkwardly but astutely captures this dialectic (Swyngedouw 1992; on the localities debates see Smith 1987; Cooke 1987; Urry 1987; Beauregard 1988; Cox and Mair 1989; Duncan and Savage 1989; for the connection between localism and postmodernism see Cooke 1990).

The broadly conceived insufficiency of a postmodernism geared to new strains of capitalism, or of a narrow localism that does little more than recoil from globalism, should not encourage the elision of the dilemmas that threw up these revolts in the first place. By the same token, the new transdisciplinary consensus around simultaneous global/local transformation now hides as much as it reveals. Especially in the US academy where a spatial sensibility is quite underdeveloped, global/local talk too quickly substitutes for substantive spatial analysis. We might even say with Mauro that such discourse entrances readers and writers with the insinuation of a replete global map that is actually quite empty. When space is everything, as with antihistoricist strands of postmodernism, it is simultaneously nothing, and Mauro's dilemma of a world uninhabited between local and global is repeated. If, as Giovanni Arrighi has it, Venice in this period is "the true prototype of the capitalist state" (Arrighi 1994: 37). Mauro's mapping would seem to combine the modernist instrumentality of a global gaze with the reflexive representational angst of a self-questioning postmodernism.

More than four centuries later, it remains a potent combination. Arrighi's *The Long Twentieth Century* is extraordinary in making the connection between the political economy of the hegemonic Italian city states during the Renaissance and the apparent crisis of capitalism today. But it is less his *longue durée* from the fifteenth century than his diagnosis of the present that concerns me here. For unlike most commentators on late twentieth-century capitalism, Arrighi deploys an astute geographical sensibility that sharpens his analysis of contemporary global change. If the longer historical sequence of four 'long centuries' of capitalist development from the fifteenth century is open to critique on a number of historical as well as theoretical grounds, he is nonetheless correct to identify an expanding territorial scale of power with each successive phase of capitalist dominance. He makes a convincing case that the period since the early 1970s marks a vital transformational crisis in world capitalism. This observation is of course not new but what is especially convincing about Arrighi's argument is the way he weaves his long historical schema of capitalist development into Marx's theory of finance capital, capital accumulation, and crisis, and his unusual sensitivity to the spatiality of capitalist development.

Dramatic industrialization of south and east Asia twinned with dein-dustrialization in Europe and North America in the 1970s marked the beginning of the crisis of a US-dominated regime of capital that had held sway since the late nineteenth century. The 'financial expansion' of the late 1970s and 1980s represented a response to that crisis and simul-taneously a second stage of it, and Arrighi queries whether subsequent financial globalization in the 1980s and 1990s marks the denouement of American hegemony. Geographically astute as it is, even Arrighi's analysis may not go far enough in highlighting the substantive and changing role of geographical space in this history of the "long twenti-eth century." Arrighi argues – again as others have done – that the old divisions between first, second, and third worlds have been ruptured to the point of obsolescence, as new Asian states take a more dominant role in the constellation of global capitals, US economic leadership is challenged, and post-cold war Russia and eastern Europe struggle toward new capitalist ambitions. But this global analysis is not matched by a corollary incisiveness about political and economic restructuring at other geographical scales and, therefore, despite the astuteness of geo-graphical argument, Arrighi misses the fact that the entire geographical framework of capital accumulation is being transformed. The very real impulses that gave rise to a new localism in England are not registered in this account, and so-called globalization is barely understandable without a simultaneous comprehension of geographical restructuring at other scales. Globalization and the remaking of the global scale are but one thread of these interconnected shifts. A whole new pattern of uneven geographical development is emerging that affects national, regional, and urban as much as global structures of social and economic production and reproduction.

In this paper I want to look only at one slice of this new pattern of uneven development. By asking the question of the relationship between a new globalism and a new urbanism, I hope not only to make some-thing of these specific interscalar connections but to illustrate the larger argument concerning the need for a theory of the production of geo-graphical scale. This attempt to tease out part of the scaled pattern within the interscalar complexity of contemporary geographical restruc-turing should further support the case for reconnecting the so-called postmodern with the modern.

Scale Bending in the 1990s

In August 1997, an impetuous Mayor Giuliani inadvertently entertained New York City with one of his periodic vents of spleen. Angry at the abandon with which UN diplomats seemed to flaunt local parking laws, and just as angry at the US State Department for its apparent capitulation to diplomatic scofflaws, Giuliani threatened to begin summary towing of diplomatic sedans, and otherwise to get tough on them. Maybe it has come to the point, Giuliani huffed, where New York City needs to have its own foreign policy.

If this threat seems whimsical, it actually expressed a very real shift in the scales of global power. In 1975 very few cities had their own global trade missions for the purpose of attracting investors and tourists and securing buyers for their products; today it is the rare medium-sized city that does not. And in the 1990s, numerous US cities and states have begun adopting their own 'urban' sanctions against foreign countries whose policies or politics they despise. In 1998, New York City and New York State, for example, threatened a ban on business with Swiss banks until the latter agreed on restitution for Nazi Holocaust survivors. As many as twenty US cities and states have implemented sanctions against Myanmar, but not Indonesia, for claimed human rights abuses. If many of these urban sanctions seem more symbolic than serious, they nonetheless caused Swiss bankers and government officials to scramble in search of a solution to the Holocaust repayment impasse. A Federal judge has ruled such foreign policy statutes by cities and states unconstitutional, but the issue is destined to reach the US Supreme Court in the first years of the twenty-first century. For his part, Mayor Giuliani has refused to join lots with other mayors and forge a global, collective, urban foreign policy. The opportunity came to his doorstep just days after his foreign policy outburst toward the UN when a conference convened in New York City under the erudite title, "Urbs et Orb": cities and the world. Organized by maverick social-democratic ex-mayor of Barcelona, Pascual Maragall, and others, the conference attendees from cities all over the world concluded that the intense pressures of globalization represented a double-edged sword for cities. On the one hand, more than at any time in the last two centuries, competitive economic and cultural pressures reached more freely and more feverishly through any skein of national state protection into the ventricles of urban economies. Yet on the other hand, precisely this erosion, however partial and uneven, of national economic power over local economies has freed cities to perform as more independent and potentially more powerful global actors. It only makes sense, the "Urbs et Orb" conference concluded, that alongside the United Nations it was now appropriate to establish a United Cities organization which would do for cities what the UN did for nations.

Such a proposal would have been fantastic in any practical sense before the 1990s, and although it still seems quixotic, its logic is nonetheless intriguing. The mayor of New York took it seriously enough to repeat the proposal while refusing to attend the conference. In a repeat of conservative US reaction to the UN, Giuliani was not about to allow his authority over a five-boroughs foreign policy to be circumscribed by adherence to any trans-urban organization. The double edge of the urban–global nexus was expressed precisely in his decision to compete against other cities rather than cooperate with the new-found promise of a collectivity of urban interests in a globalized world.

Other highly significant scale-bending events became almost commonplace in the late 1990s. Media capitalist Ted Turner announced in 1997 that in light of the UN's financial plight, caused in no small part by US congressional refusal to pay its dues, he as a private citizen was donating a billion dollars to that organization. At the same time the billionaire financier George Soros responded to the dire economic situation in Russia by providing a personal loan of half a billion dollars to the Yeltsin government, five times the level of US government's annual loans to Russia up to that point. And the Disney Company, about to release a controversial movie championing the feudal, religious monarchy of Tibet against Chinese military brutality, worried about official Chinese reaction, appointed Henry Kissinger as Disney's ambassador to China. The Magic Kingdom meets the Middle Kingdom. For Disney, concludes Cindi Katz, "a billion consumers is, after all, a terrible thing to waste" (Katz 1997: see also Zhao 1997).

These scale-bending events – cities threatening their own foreign policies, private citizens funding whole countries or even the UN, corporations hiring ambassadors to national states – suggest the profundity of the current *geographical* restructuring. That such events jar our sense of scalar propriety – every social function at its proper spatial scale – only highlights the largely hidden work that geographical scale does in ordering and maintaining our assumptions of sociopolitical normality. Today, the gamut of our postwar twentieth-century assumptions legislating the appropriate behaviors and functions of specific actors – state, corporate, urban, private – at specific scales are dissolving and new scalar assumptions are being formed. At the heart of this shift lies a restructuring of geographical scale *per se*, the implications of which are only slowly becoming evident.

Let me make the overall argument here in capsule form. In the first place, far from simply a natural given or a conceptual convenience, geographical scale also and most vitally represents a socially constructed framework for organizing political, economic, and cultural activity. Scale is the primary metric via which the flow of social differences is continually etched and re-etched *as spatial difference* in the geographical landscape. Second, to the extent that such social difference is fixed in the material landscape, however temporarily, specific social assumptions become naturalized as part of the geography of production and reproduction. Produced geographies legislate, with greater or lesser success, specific assumptions about social and political difference and right. Third, then, to the extent that scale differences embody a repository of presumed social difference, the critique of received geographies is simultaneously and necessarily a recovery of politics. To the extent that scale is the most crystallized spatial repository of social difference, an "archaeology of scale" (to use the Foucaldian language) reveals a politics of social difference.

Scale Theory

In the 1980s and early 1990s, it gradually became apparent that a restructuring of scale played an integral if rarely perceived part in the global transformation that Arrighi highlights (Taylor 1982; Smith 1984; Herod 1991; Swyngedouw 1992; Smith 1992a; Smith 1992b; Smith and Dennis 1987). This became increasingly evident in the 1990s as greater attention was paid to different aspects of this restructuring of scale. From its roots in political economy, the scale discussions broadened significantly. Regarding the shifting roles of cities *vis-à-vis* national states and global political economy, for example, a number of approaches have since been made spanning questions of new identity production, urban citizenship, and the political forging of localities (Holston and Appadurai 1996; Keil, Wekerle, and Bell 1996; McDonogh 1997; Öncü and Weyland 1997). And yet much of this discussion continues to be framed by arguments about globalization. Adherents of what we might call 'the strong globalization thesis' maintain that there has been a decisive scalar shift in the locus of power from the national to the global scale. National borders are significantly eroded as barriers to economic, cultural, or social mobility, as well as political power, and this gives rise to a virtually borderless world where the prerogatives of national power are now severely circumscribed. In one version, globalization heralds 'the end of the nation state' (see among others, Ohmae 1990, 1995). This may be a matter of jubilation, as in the halls of financial capital, or of bitter resignation, as in much of southeast Asia where financial globalization opened the region up to the 1997 crash – or in Iraq, where globalization in the 1990s combined economic sanctions with military suppression associated with a 'new world order.' Of course, globalization may be neither as complete nor all-consuming as this position suggests. Some of the same economic, political, and cultural forces associated with globalization may actually have strengthened certain national and regional boundaries and encouraged the emergence of new nationalisms; from central Africa to the Balkans this has resuscitated old place-based social and political localisms in particularly ugly forms. The brutality of the latter notwithstanding, local resistance and local political structures can actually be strengthened. A reassertion of the local actually accompanies whatever passes for globalization (see, for example, Cox 1997). The fundamentalist Christian right in the United States and the left wing of the Labour Party in Britain prior to 1997 shared very little, one assumes, but they both held that whatever the merits of thinking globally, acting locally was the fulcrum of political activism. Against these starkly posed alternatives, the pressing need to rethink a more sophisticated treatment of global *cum* local change prompts the more general, framing question of where geographical scales come from in the first place, how they change, and what work they do. In one of the most sophisticated attempts to integrate global and local change in an overarching theoretical and political framework, Eric Swyngedouw insists that we conceive of globalization and localization as parallel and simultaneous processes. So definitive is the production of scale, he suggests, that it may even be necessary to jettison the overly static scalar language of 'global' and 'local' in favor of a much more rigorously worked out dynamics of geographical scale (Swyngedouw 1997: 142; Swyngedouw 1992; see also Brenner 1998). Hence the coining of 'glocalization.'

Geographical scale results from the territorialization of highly fluid social relations and differences. Scale is always a temporary fixation in the geographical landscape, of course, enduring for a shorter or longer term. Whatever the preexisting national state forms, the scale of the nation-state, for example, has been of relatively short duration, dominating global history for little more than two centuries; in postcolonial Africa and Asia, its history is more correctly measured in decades. More generally, the establishment of geographical scale functions to mediate, spatialize, and thereby displace the socioeconomic contradiction between competition and cooperation. The placement of socially sanctioned boundaries around the body or the nation-state, the urban or the global, mobilizes a logic of scale that already embodies certain assumptions about identity and difference (Smith 1984; Swyngedouw 1992: 43–4; Smith 1992a). Inclusion within a group defined in specific scalar terms implies a conditional agreement to cooperate with others included in the group while competing – along some economic, cultural, or social lines – with those of other groups. The territorialization of these social relations can be haphazard, but in societies where economic exchange is routine through the market and political relations are mediated through relatively stable state forms, systems of socially produced geographical scale increasingly organize and mediate the territorialization of identity and difference. Specific national states and clan compounds, economic regions and city states all express not only specific social relations at a given time and place, but systems of existing or emerging scales of social interaction.

Geographers and others have long conceived spatial organization as a force of social production. Territorial infrastructure is constructed as a vital organizational landscape to facilitate social production and reproduction. But to the extent that this is true, it also makes sense to conceive of geographical *scale* as an organizational metric for the production of space (Swyngedouw 1992; on space as a force of production see Buch-Hansen and Nielson 1977; Lefebvre 1991; Santos 1994). Scale is to the production of space what private property is to capitalism. Geographical scale is simultaneously a means of political containment and empowerment. To the extent that social struggles and revolts, labor contracts, and place-bound allegiances can be contained at specific scales and thereby made competitive with commensurate struggles and allegiances in other places, scale becomes a political as much as an economic technology, a means of social control inscribed in the territorial organization of the landscape. But the converse is also true. To the extent that specific places, framed by existing scalar relations, become the basis for struggles and organization opposed to ruling classes and elites, existing political assumptions and economic structures, geographical scale becomes a means of social and political empowerment. As Swyngedouw puts it: "Sociospatial struggle and political strategizing, therefore, often revolve around scale issues, and shifting balances of power are often associated with a profound rearticulation of scales" or the production of new scales (Swyngedouw 1997: 170; Herod 1991, 1998; Smith 1992b).

Geographical scales are organized hierarchically. Much as we may wish to contest social hierarchies, wishing them away is no substitute for developing the political means by which such hierarchies of power can be challenged, even dissolved. But it would be just as misguided to conceive of the hierarchy of geographical scales as possessing a rigid architecture. First, the territorial organization of scales necessarily responds to changes in the economic, political, and cultural conditions out of which they pupate. Second, if geographical scales are connected in a nested hierarchy, they are equally concurrent and coincident if not necessarily synchronous. An event or string of events happening in one discrete place may or may not register as simultaneously of national or global significance. Thus the reality of a scalar hierarchy of sociospatial power at any given moment in no way refutes the plasticity of scalar relations or the structured contingency of social and political conditions and relations amidst which scalar significance is assigned. Why, to take one obvious example, did the 1998 revolt against the Indonesian ruling class, symbolized especially by Suharto, have such immediate global significance when the independence struggles of the East Timorese, brutally suppressed at the cost of many more lives, were for 24 years widely treated as merely regional?

The answer is not at all geographical but lies in the connections between struggles for national and subnational identities, the predicament of the global economy and the interests of its major players, and power over the media and military. Yet assumptions of geographical hierarchy stand available as means of contextualizing and naturalizing one struggle while highlighting another as critical.

The scalar significance of particular events is certainly delineated, in the first place, in relation to existing scalar hierarchies, but this also occurs within the context of real or potential political struggles. If we shift perspective now and take a stand not on a distanced vision of geographical hierarchies but on the ground of these political struggles, geographical scale becomes a vehicle and strategy of political empowerment. To the extent that struggles can be contained at one scale it becomes imperative to 'jump scales' as a means of avoiding containment, and the jumping of scales becomes the means by which new scalar arrangements are forged (Smith 1992b: 66).

We therefore have a triple result. Scale is a social technology, but insofar as it is also a political repository of assumed social difference and rights, it is simultaneously a deeply disguised ideological expression – 'representational site' – of ruling social ideas. To the extent that this inherent representationality of scalar hierarchies is revealed via political, economic, and cultural critique, geographical scale is transformed into a force of alternative political production, a resource for social and political revolt.

The New Urbanism

The mention of new urbanism inside the covers of a book on 'postmodern geographical praxis' will for many connote the upstart, antisuburban school of planning, architecture, and design that adopted this name in the 1990s. But it is not to this particular postmodern modernism, more accurately referred to as neotraditionalism, that I refer here. Rather, the point of appropriating this title is to highlight the fact that the contemporary restructuring of spatial scale glossed by reference to 'globalization,' is equally responsible for initiating a new urban regime – a new urbanism – featuring a dramatically restructured relationship between economic production, social reproduction, and political governance. Not that *this* new urbanism is wholly unrelated to the design tradition of the same name. In fact, a plausible argument can be made that in addition to an aesthetic rejection of suburbia, architectural new urbanism represents a highly localist political reaction against the new urban realities of city life.[1]

The most straightforward implication of the globalization rhetoric is that an economy previously organized into different national units is now increasingly global. The scale question is of course much more complicated than this. The *global* scale was already well in place before the 1980s, having crystallized throughout the century as the scale of capital circulation and the market.[2] The *national* scale, a project of more than two centuries and many bourgeoisies, was largely in place by the beginning of the century but was completed with the decolonization of Asia and Africa between the late 1940s and the early 1990s. The nation-state emerged as a fixed geographical form via which to regulate economic competition in the market and the contradictory impulses of equalization and differentiation of levels and conditions of production. Nation-states came into the world twinned with national capitals. Within them, relatively coherent and discrete regions emerged as the production platforms of national economies.

Cities have historically performed multiple functions ranging from the military and religious to the political and commercial, and as Arrighi's discussion of Venice and Genoa in the fifteenth and sixteenth centuries reminds us, military, political, and territorial power has at various places and times been vested in the urban scale. If the expanded scale of economic accumulation rendered city states obsolete in the transition to capitalism, the loss of *political* power at this scale hardly made cities themselves obsolete. Quite the opposite; the percentage of the world's population living in cities was about to soar. By the late nineteenth and through most of the twentieth century, cities increasingly if unevenly came to be defined according to their role in the organization of social reproduction – the provision and maintenance of a working-class population. The city was also an extraordinary centralization of means of production and productive activity, cultural work, and political strife – and much more; but as soon as the division of labor between social production and reproduction became simultaneously a spatial division, the scale of the *modern* city was defined in terms of the limits of the daily movement of workers between home and work. We could fill out this picture of scalar divisions by conceptualizing the local community as the scale of physical reproduction, the home as the locus of biological reproduction, and the body as the scale at which social identities are primarily crystallized.

The Keynesian city of postwar advanced capitalism represented the zenith of this definitive relationship between the city and social reproduction. This is a consistent theme running through the work of major European and American urban theorists of the last third of the century (Lefebvre 1971; Harvey 1985; Castells 1977). The Keynesian city was, in many respects, the welfare hall for each national capital, combining the functions of social support and the reproduction of a national labor market. Indeed, the so-called urban crisis of the late 1960s and 1970s was widely interpreted as a crisis of social reproduction, having to do with the dysfunctionality of racism and patriarchy and the contradictions between an urban form elicited according to strict criteria of profitability yet one that had to be justified in terms of the efficiency and efficacy of social reproduction. But by the 1970s, the idea of separate 'national economies' was obsolete. International linkages *per se* were not new, of course, but the level of economic integration across national boundaries *was*, and with commercial and financial capital largely globalized already, it remained only for production capital also to 'jump scales.' This was the real basis for the 'discovery' of globalization by the late 1980s, a process that was dramatically fuelled if not caused by the generalization of computer technology, the comparative cheapening of air transportation, and the parallel cheapening of most raw materials.

The erosion of the national scale, at least in economic terms, is having a direct and rapid impact on cities, and to understand how and why it is necessary to identify the different dimensions of the weakened national scale. First and most obviously, enhanced communications and deregulation have increased capital mobility and dramatically expanded the range of capitals that are free to move where lower costs of social reproduction in turn lower the costs of production. Second, unprecedented labor migration flows in the last quarter of the century have increasingly distanced local economies from automatic dependency on home-grown labor. There may not yet be a single global labor market but there are many international labor markets all of which are globally interdependent (Castells 1996: 239). Third, forced into a more competitive mode *vis-à-vis* capital and labor, local states (including city governments) have offered carrots to capital and applied the stick to labor; they have become much more selective about the extent and level of subsidy for social reproduction since they can depend to a greater extent on imported labor whose costs of reproduction are borne elsewhere. Fourth, the same pressures applied to the national state have led to a dramatic erosion of social capital provision at that scale also, intensifying the pressure on city governments to further sever their responsibilities for social reproduction. Finally, amidst the restructuring of production beginning in the 1970s and with class- and race-based struggles broadly receding, city governments had an increased incentive to abandon that sector of the population surplussed by both the restructuring of the economy and the gutting of social services. Comparatively low levels of struggle were crucial in the wholly military response by government to the Los Angeles uprisings after 1992, a dramatic contrast with the ameliorative response following the Kerner Commission Report in the late 1960s.

Two mutually reinforcing shifts have consequently restructured the functions and active roles of cities. In the first place, systems of production previously defined at the regional scale were less and less defined in terms of the needs of national capitals, contributing in the 1970s and 1980s to the rash of deindustrialization in regions from the Ruhr to the US Midwest while whole new regional ensembles of production emerged. Unlike the disintegrating industrial regions of Europe and North America that they replaced – and even when that replacement appeared to take place within the shells of the old regions – the new regions emerging in the late twentieth century are from the start defined in terms of the global market; they are first and foremost engines of global rather than national economies. Crucially, they also involve a significant shift in the established scale hierarchies and specifically in the scale at which the new production complexes are organized. Disappearing are the extensive manufacturing regions that might extend for hundreds of thousands of square kilometers and envelop numerous specialized urban manufacturing centers. In their place have come production 'regions' that are little more than extended metropolitan centers, or regions strung between such centers. Silicon Valley and Shanghai, Bangkok and São Paulo to a considerable extent succeed and supplant the American northeast and the English north and midlands as the geographical shape of global production. The metropolitan world comes to dominate the regional world rather than the other way round. In retrospect, the intense focus by economic geographers on regional industrial systems and relations in the 1970s and 1980s is symptomatic. Modern economic regionalism had its zenith in this period and the remaking of twenty-first century metropolises as leading platforms of global production is hollowing out inherited regional structures.

But the corollary is also taking place. Confronting staccato economic crises between the early 1970s and the 1990s, punctuated by bouts of rapid market expansion, national states found themselves increasingly unhinged from the economies of their territories and, given the extent of capital and labor migration, no longer necessarily responsible for bearing the costs of social reproduction associated with the labor forces occupying the state's territory. In the United States, President Carter's attempted urban plan of 1978 represented the last time that an attempt

was made to tie the economic fate of the country's cities so definitively
to a national Keynesian economy. The failure of that plan, which left
the fiscal crises of many older cities unresolved by federal intervention,
gave the first intimation of a delinking between urban and national
economies. The demise of liberal urban policy followed piecemeal until
President Clinton's cynical guillotining of the social welfare system in
1996. The national state's relinquishment of significant responsibility
for social reproduction and its devolution not to cities but to the states,
not only broke the welfare nexus between cities and national states
but released the cities themselves from administering the welfare arm of
social reproduction. Not by accident, cities replaced welfare with '*workfare*,' underlining the transformed role of cities in the world economy.
And not by accident is a similar shift from welfare to workfare underway in the majority of so-called advanced capitalist states.

If the demise of liberal urban and welfare policies has to be understood as in part a rational (market-based) response to globalized economic competition on the one hand and increasing independence from
domestic (urban-centered) social reproduction on the other, it is equally
an abrupt reassertion of class power, and of race, and gender power
expressed through the pores of class. The restructuring of geographical scale represents a class, race, and gender specific remapping of
the known world. Since the eighteenth century at least, the national
economy provided the arena in which capitalist economic and political
power was expressed, but this equation of power and space no longer
holds at the beginning of the twenty-first century.

The profundity of this shift is difficult to exaggerate and, as Bill Readings has suggested, penetrates deep into our fragmenting 'national cultures.' For him, the crisis of the nation-state crystallizes directly in a
crisis of cultural reproduction, spawning not just the corporatization of
the university – historically trusted with the task of reproducing national
cultures – but the so-called 'culture wars' and indeed cultural studies
themselves. As the erosion of national power provokes a 'delegitimiza-
tion of culture,' cultural studies, for Readings, represents a desperate
attempt to reconstruct a cultural vision as an all-encompassing but
'dereferentialized' gestalt of global corporatism. The cultural turn represents a thoroughly Enlightenment move at a higher scale, a "'postmodernity' of the university" keyed – as minor to major? (Katz 1996)
– to the corporatization of knowledge. Insofar as culture is posited as
everything, it is rendered a fetish, and cultural studies vigorously abets
the commodification of 'culture' (Readings 1996).

That the economic rationale for nation-states is significantly eroded
on the cusp of the twenty-first century is surely indisputable. One might
even be tempted to apply Jürgen Habermas's felicitous phrase to these

quintessential geographical constructs of modernity, that they are destined to become dominant but dead. Or, as political geographer Peter Taylor (1995: 58) has put it, "cities are replacing states in the construction of social identities." But as this suggests, power is not exhausted by the economy; far from it, and it would be a mistake to confuse the erosion of economic power at the national scale with a generalized powerlessness. Again the lesson from Venice and Genoa is salutary. When city states gradually relinquished political economic power to nation-states, this did not mark the "end of the city." Rather, cities themselves, and the urban scale *in toto*, were restructured. Thus a substitution and reworking of the political and cultural for economic rationales may equally transform the role of nation-states today.

The new urbanism, then, would seem to promise a reappropriation of the rationale enjoyed by cities in the earliest days of industrial capitalism in Europe, a time when, not incidentally, the economic authority of the nation-state over cities was also weak. But the leaders in this combined restructuring of urban scale and function lie not in the old cities of advanced capitalism where the disintegration of traditional production-based regions and the increasing dislocation of social reproduction from the urban scale is partial, painful, and unlikely to pass unopposed. In many cities of Asia, Latin America, and Africa, however, where the Keynesian welfare state was never installed, and the definitive link between the city and social reproduction was never paramount, the fetter of old forms, structures, and landscapes will be much less. These metropolitan economies will operate as the new production platforms of a new globalism. Unlike the suburbanization of the postwar years in North America and Europe, dramatic urban expansion of the early twenty-first century will again be led by the expansion of social production rather than reproduction. In this respect at least, Lefebvre's announcement of an urban revolution redefining the city and urban struggles in terms of social reproduction, or indeed Castells' definition of the urban in terms of collective consumption, will fade into historical memory.

If in the Keynesian world the geographical sprawl of suburbs in western cities could potentially undermine the law of value insofar as the trade-off between lower house prices and higher-cost commutes destabilized the value of labor power – and thereby contributed to the formation of economic crises – the argument is increasingly transposed in the twenty-first century city. The production of metropolitan space still potentially constrains the law of value except that it is more and more the environmentally inspired deterrence of production rather than social reproduction that will press against the law of value. Some of the constraint may be similar; traffic jams, long commutes and exorbitant rents now characterize São Paulo and Bombay as much as London and New York. But with highly decentralized, interlinked production nodes nestled around metropolitan areas, rather than a single concentrated center, the effects of these constraints are transformed. More broadly, the much acclaimed postindustrial city increasingly appears in this light as a quixotically narrow and privileged self-delusion belonging to a specific time and space. Cities like Shanghai and Lagos, Mexico City and Manila will challenge the more traditional urban centers, not just in size – they have already done that – but, primarily as leading incubators in the global economy, progenitors of new urban form, process, and identity. They are already cities in which the journey to work has become excessive for many people without obviously challenging the law of value.

Conclusion

As he stares at his completed map, Frà Mauro gradually morphs into his own cartographic creation. Having lived a 'conspiracy of escape' throughout his life, welcoming strangers to tell their tales inside the monastery walls, he comes to realize the intolerable responsibility he has placed on himself by mapping the world and thus changing it. The only escape from his own dangerous conspiracy, he now sees, is to re-merge with what he has made. Overcome by an overpowering lassitude, he spills into his map from the margins and spills over into the "most mysterious principality of them all – that of the kingdom of 'no-knowledge'" (Cowan 1996: 146–8). Emptiness, death as sublime redemption.

The Christianity of Mauro's journey is no accident, of course. As Cowan notes, he "began to recognize the power of invisible events to change the course of history." An 'invisible geography' came to affect how he thought about place. In the spirit of Mauro's idealism (matter emptied of its content), Cowan interprets this as a purely spiritual, imaginative geography (Cowan 1996: 151), but it seems to me that it is precisely his idealism, the possibility of an identity morph from individual to world, that forces on Mauro the recognition that his lifelong escape was intolerable and dangerous, and must itself be escaped. Rather than escape the monastery, he escapes into and therefore from the world he has constructed.

Theories of geographical scale also represent an attempt to decipher 'invisible geographies,' but that project need not lead to the serene nihilism of Mauro's end. Too late he seems to have recognized that matter emptied of its content represented a suicidal road to knowledge. An archaeology of scale, a theory of the social production of geographical scale, is precisely about refilling the 'content' of social space with its ideologically 'emptied matter.' Insofar as scale is the socially devised metric of spatial differentiation, it is difficult to conceive of a geographical project that is more acutely political. If this quest is entrancing along the way, so much the better, as long as we do not replicate the self-destructive mistake derived from Mauro's isolation and confuse ourselves for the world.

NOTES

1   On the design school, see P. Katz (1994). For a critique, see Rutheiser (1997). For an extended discussion contrasting contemporary urban restructuring with the new urbanist design tradition, see my "Which New Urbanism: New York in the Revanchist 90s," in R. Beauregard and S. Watson, eds., *Urban Moments*. Sage: Beverly Hills, forthcoming.
2   The following several paragraphs are drawn from my "Retro Modern or Revolutionary: Scale Shifts and Political Reaction in Twenty-First Century Urbanism," *City and Society*, Annual Review 1997: 37–41.

REFERENCES

Arrighi, G. (1994). *The Long Twentieth Century*. London: Verso.
Beauregard, R. A. (1988). In the Absence of Practice: The Locality Research Debate. *Antipode* 20: 52–9.
Brenner, N. (1998). Global Cities, Glocal States: Global City Formation and State Territorial Restructuring in Contemporary Europe. *Review of International Political Economy* 5: 1–37.
M. Buch-Hansen and B. Nielson (1977). Marxist Geography and the Concept of Territorial Structure. *Antipode* 9(2):1–12.
Castells, M. (1977). *The Urban Question*. London: Arnold.
Castells, M. (1996). *The Rise of Network Society*. Oxford: Blackwell.
Cooke, P. (1987). Clinical Inference and Geographic Theory. *Antipode* 19: 69–78.
Cooke, P. (1990). *Back to the Future: Modernity, Postmodernism and Locality*. London: Unwin Hyman.
Cowan, J. (1996). *A Mapmaker's Dream: The Meditations of Fra Mauro, Cartographer to the Court of Venice*. Boston: Shambhala.
Cox, K. and A. Mair (1989). Levels of Abstraction in Locality Studies. *Antipode* 21: 121–32.
Cox, K., ed. (1997). *Spaces of Globalization: Reasserting the Power of the Local*. New York: Guilford.
Duncan, S. and M. Savage (1989). Space, Scale and Locality. *Antipode* 21: 179–206.

Harvey, D. (1985). *The Urbanization of Capital.* Oxford: Blackwell.

Herod, A. (1991). The Production of Scale in United States Labor Relations. *Area* 23: 82–8.

Herod, A. (1998). *Organizing the Landscape: Geographical Perspectives on Labor Unionism.* Minneapolis and London: University of Minnesota Press.

Holston, J. and A. Appadurai (1996). Cities and Citizenship. *Public Culture* 8: 187–205.

Katz, C. (1996). Toward Minor Theory. *Environment and Planning D: Society and Space* 14: 487–99.

Katz, C. (1997). Power, Space and Terror. Paper presented at the University of California, Berkeley, Oct.

Katz, P., ed. (1994). *The New Urbanism.* New York: McGraw Hill.

Keil, R., G. Wekerle, and D. Bell, eds. (1996). *Local Places in the Age of the Global City.* St. Paul, Minn.: Black Rose Press.

Lefebvre, H. (1971). *La Révolution urbaine.* Paris: Gallimard.

Lefebvre, H. (1991). *The Production of Space.* Oxford: Blackwell.

McDonogh, G. W. (1997). Citizenship, Locality and Resistance. *City and Society Annual Review*: 5–33.

Ohmae, K. (1990). *The Borderless World: Power and Strategy in the Interlinked Economy.* New York: Harper Collins.

Ohmae, K. (1995). *The End of the Nation State: The Rise of Regional Economies.* New York: The Free Press.

Öncü, A. and P. Weyland, eds. (1997). *Space, Culture and Power. New Identities in Globalizing Cities.* London: Zed Books.

Readings, B. (1996). *The University in Ruins.* Cambridge, Mass.: Harvard University Press.

Rutheiser, C. (1997). Beyond the Radiant Garden City Beautiful: Notes on the New Urbanism. *City and Society Annual Review*: 117–33.

Santos, M. (1994). *Técnica, Espaço, Tempo: Globalização e meio técnico-científico informacional.* São Paulo: Editora Hucitec.

Smith, N. (1984). *Uneven Development: Nature, Capital and the Production of Space.* Oxford: Blackwell.

Smith, N. (1987). Dangers of the Empirical Turn: The CURS Initiative. *Antipode* 19: 59–68.

Smith, N. (1992a). Contours of a Spatialized Politics: Homeless Vehicles and the Production of Geographical Space. *Social Text* 33: 54–81.

Smith, N. (1992b). Geography, Difference and the Politics of Scale. In *Postmodernism and the Social Sciences*, eds. J. Doherty, E. Graham, and M. Malek. London: Macmillan, 57–79.

Smith, N. (1997). Retro Modern or Revolutionary: Scale Shifts and Political Reaction in Twenty-First Century Urbanism. *City and Society Annual Review*: 37–41.

Smith, N. (forthcoming). Which New Urbanism: New York in the Revanchist 90s. In *Urban Moments*, eds. R. Beauregard and S. Watson. Beverly Hills: Sage.

Smith, N. and D. Ward (1987). The Restructuring of Geographical Scale: The

Coalescence and Fragmentation of the Northern Core Region. *Economic Geography* 63: 160–82.

Swyngedouw, E. (1992). The Mammon Quest: 'Glocalisation,' Interspatial Competition and the Monetary Order. In *Cities and Regions in the New Europe*, eds. M. Dunford and G. Kafkalas. London: Bellhaven Press, 39–67.

Swyngedouw, E. (1997). Neither Global nor Local: 'Glocalization' and the Politics of Scale. In *Spaces of Globalization*, ed. K. Cox. London: Guilford, 137–66.

Taylor, P. (1982). A Materialist Framework for Political Geography. *Transactions of the Institute of British Geographers* 7: 15–34.

Taylor, P. (1995). World Cities and Territorial States: The Rise and Fall of Their Mutuality. In *World Cities in a World System*, eds. P. Knox and P. Taylor. Cambridge: Cambridge University Press, 42–68.

Urry, J. (1987). Society, Space and Locality. *Environment and Planning D: Society and Space* 5: 435–44.

Zhao, B. (1997). Consumerism, Confucianism, Communism: Making Sense of China Today. *New Left Review* 222: 43–59.

Mappings . . .

8

Millennial Geographics

Denis Cosgrove and Luciana L. Martins

Celebration of the millennium is an act of creative globalization which universalizes the Western, Christian space-time that underpins a largely secular geographical imagination. In order to be enacted, the meanings of millennialism – death and renewal, origins and ends, memory and desire – have not only to be attached to a specific calendrical moment, but also to be mobilized at particular locations across the globe. This chapter focuses primarily on millennial celebrations in two European capital cities, Rome and London. These celebrations seek to universalize complex meanings, memories, and desires through concrete interventions: buildings, spectacles, and performances, and to animate historic *genius loci* in determinate locations within a contemporary global context. Although connected to a long heritage of universalizing consumption spectacles within the modern city, millennial activities might be regarded as postmodern in both the geographies they seek to stage and in their modes of spatial realization. They may also be described as 'mappings' in the sense that has become familiar since Surrealist psychogeographers introduced the idea of performance in urban space during the 1950s (Debord 1955; Pinder 1996). While lacking the subversive intention of those projects, millennial celebrations measure, trace, and represent spatio-temporal concepts, experiences, and connections.

This discussion uses millennial celebrations in London and Rome to reflect upon two theoretical concerns within contemporary geography. First, we examine how millennial mappings face the contradictions between the universalism and implicit time–place fixity of the idea of millennium on the one hand, and contemporary relational time–space thinking and polyvocality on the other. Late twentieth-century geography concentrates on a relational rather than a fixed concept of space, conceptualizing space as a sphere of multiple interrelationships, always in the process of becoming (Massey 1999: 283). Such a non-essentialist notion of space introduces our second theoretical concern: the idea of 'performative mapping.' By this term, we refer to the ways in which *genius loci*, the capacity of particular places to communicate intense and unique meaning across space and time, is actively made and remade. Argyro Loukaki's recent critical examination of *genius loci* highlights both the contradictions of universalism and localism and the social contestations surrounding collective meaning and memory fixed in any location (Loukaki 1997: 308–10). She also points out the role of creative artists in 'translating' their creative inner life as part of social reality and collective experience. Performative mapping implies the contemporary convergence between artistic and scholarly activity in geography, simultaneously understanding and activating spatialities, which is the subject of the second part of this chapter. Such convergence moves beyond sterile science-versus-humanities debates within geography, reflecting much broader epistemological shifts apparent at the millennium. Developments in the study of complex systems signal a shift from "the dominant science of quantities that has characterised modernity, towards a science of qualities that is emerging in the post-modern era" (Goodwin 1997: 118). This implies an active engagement in the processes of knowledge-making in which we are immersed, a movement that not only historicizes our understanding of scientific practices (Smith 1998: 273), but also emphasizes the creative and imaginative dimensions of all enquiry. The boundaries between once distinct categories of activity – the scholarly and artistic, for example – are thus becoming less clear. Creative artists increasingly draw upon, and contribute to, the practice of geography, while geographers themselves are adopting more creatively imaginative ways of making and communicating knowledge. Our critical reading of millennial projects in Rome and London thus opens with their common significance as post-imperial cities whose deep-rooted engagements with eurocentric and modernist mappings are inscribed into their physical landscapes. We recognize that each has a particular claim to a *genius loci* connected to the millennium, and proceed to consider how these claims are variously activated.[1]

The Calendar and the Map

In its emphasis on the figure of a thousand (*mille*), millennialism naturalizes a language of universal measure that is historically and culturally specific to a global and implicitly imperial Western time–space mapping dating back to Augustus Caesar. The significance of the millennial moment (00:00 h, January 1, 2000) stems more directly from a papal decision in 1582 to reform the Christian calendar. Commercial and media hype about this date is built upon a recognition that, however arbitrary they may actually be, such calendrical moments are invested with significance as breaks in the flow of social time.[2] The point of passage between millennia is freighted with profound significance, connected to teleological beliefs and imaginings about social and personal origins, about renewal, and about ends. Cultural responses to the passage of the first Christian millennium in medieval Europe were overwhelmingly apocalyptic. Despite the secular character of modern societies, current intellectual and cultural debates over the values of modernity and 'progress,' the profound effects of information technology on modalities of human experience (Baudrillard 1998), and such varied social phenomena as 'New Age' irrationalism, religious fundamentalism, and syncretism, all suggest that, while different in language, the social responses to the passage of the second millennium are just as intense.

Calendar dates, however, are not merely markers of time: they originate and are expressed in measures of space (the movement of the planets, the decay of radioactivity in materials, the distance between arboreal growth rings). And they 'take place,' they are marked locally across the earth's surface by both permanent structures and performances. The calendar and the map are thus united in recording such cosmic events and their meanings within a geography whose most characteristic conventional icons are those of the cosmos, the globe, and the whole earth. More local expressions may be found in the symbolic spaces, monuments, and rituals staged within cities, signifying, in Western culture, the historic connection between the idea of a 'world city' and the globe (urbs et orbis). The Tower of the Winds in the Roman forum at Athens, or the medieval European convention of marking the edges of the world maps with figures representing both the compass points and the trumpeting angels of Apocalypse, provide examples.[3] Recent work on the history of cosmographic mapping and global images from the Renaissance to the present has explored some of these relations, showing that they are as potent in the contemporary world as in the past (Cosgrove 1994; 2000). In a study of the Globe structure built in Stockholm in the late 1980s, Alan Pred (1995: 181–98) not only places this 'spectacular space' in a two-century tradition of great globes designed to universalize 'industrial modernity, high modernity and hypermodernity,' but also points to the coincidence of the project with the emerging globalization of capital and more local changes in the political economy of Sweden and its capital city. He notes the "register of post-modern devices – pastiche, irony, eclecticism, the appropriation of 'local history' and 'collective memory', the representation of local architectural icons, the more general 'citing' of historical forms and (near schizophrenic) double coding" whereby the Globe is, in Pred's opinion, intended "to aesthetically pacify with powerful conflict-denying façades while simultaneously hiding (attempts at) a new joint consolidation of economic and social power" (Pred 1995: 187, 194, 195). But Pred does not trace either the deeper historical roots of such urban spectacles, which in fact long predate modernity, nor reveal the precise connections between the *form* of the Globe and the contextual processes in which he locates the Stockholm project. Those deeper and more complex cultural connections are perhaps more explicitly revealed in millennial celebrations.

At the end of the first millennium, there was a widespread belief in Christian Europe that Christ's second coming was imminent and that it would be geographically located: the transfigured Christ would appear from the West, the direction not only of the setting sun, but also, as then unknown oceanic space, the location of Europe's most intense desires and fears. Today, some of the anxieties surrounding the millennium are equally apocalyptic. One example was found in the concerns over the global impact of the Millennial computer bug (Fallows 1999). Appropriately, in the context of a postmodern spatiality, it was predicted that its impact would not emanate from a single point, but would rather spread out through the networks of virtual space, affecting hospital energy sources, food provisions, traffic control, and airlines, for example.

Nonetheless, there is a parallel sense of millennial renewal and futurist speculation similar to that associated with *fin-de-siècle* modernity. From at least the mid-nineteenth century, as Pred and others have shown (Pred 1995; Greenhalgh 1988; Bennett 1996; Mattelart 1996), ephemeral landscapes designed to encompass the world have periodically been constructed in virtually all of the major Western capitals, and many other European and American cities, in order to celebrate the achievements of a supposedly universalized humankind. Of all the moral justifications for expenditure on such projects, it is the belief in progress, in new technology and the onward march of civilization, which have been the most prominent. 'Peace among nations' has been the common refrain of universal exhibitions, Olympic Games, and other global events during the twentieth century, though a fundamental asymmetry between the exhibitors and the exhibited has been a characteristic feature of all these spectacles (Mattelart 1996: 131–2). The content and ideological significance of world exhibitions have of course varied according to location and historical period. If the imperial theme was a constant in British and French exhibitions, for example, it was not so prominent in America. Nevertheless, one assumption has been constant: the westward-pointing arrow of civilization's progress has signalled the principal temporality to be celebrated. By contrast, the millennium is an instant of stasis, a point in time when past and future, memory, and desire, coincide. In the context of the "reversal of the whole of our modernity's relation to time" which Jean Baudrillard (1998) portrays as a feature of postmodernity, it is the countdown – the fear of ending and of closure – that is emphasized, while the image of a unifying and progressive West is increasingly acknowledged to be redundant.

What is true of Western conventions of temporality is equally true of its traditional global spatial imaginary. The mapped globe which emerged over the course of a half-millennium between 1450 and 1950,[4] inscribed with its linear fixities of latitude and longitude, of continental coastlines and of political territories, has been displaced by the blurred surfaces and relativities of satellite images of earth, the interconnections of virtual global hyperspace, and the permeable territorialities of a decentered, postcolonial sphere. In such a fluid and uncanny space–time, attempts to 'map' the millennial moment in specific locations acquire considerable poignancy. As Terry Eagleton (1999) has recently pointed out, geography "can question postmodern assumptions as much as underpin them," insofar as practices of mapping are sustained by some degree of epistemological realism (even acknowledging their social-constructed nature, maps nevertheless imply representation of a world beyond themselves), contrasting sharply with the postmodern episte-mological embrace of positioned constructivism, heterotopic difference, and dogmatic anti-universalism.

Millennial Cities: London and Rome

In 1884, at perhaps the high point of the history of world exhibitions, the universal Prime Meridian was established at Greenwich. By this act, London proclaimed its centrality in the measurement of secular time and the representation of global space.[5] In the year 2000, therefore, Green-wich is the site for major millennial celebrations, symbolized above all by its controversial Millennium Dome, a 320-meter diameter, 50-meter high, circular structure with a circumference of one kilometer, sus-pended like a convex clock face from twelve 100-meter steel masts. Although not a dome in the exact sense (i.e. a hemisphere), its name and form invoke earlier cosmic structures in the Western tradition, going back to the Pantheon in Rome. Meanwhile, in Rome itself, 2000 is the year of the bimillennial *Giubileo* (Jubilee), an event which will attract more than 30 million Catholics from across the globe,[6] and whose centerpiece is Michelangelo's domed basilica of St. Peter's, the navel of Catholicism. Despite various modernist attempts to recalculate the calendar from a 'Year Zero,' associated especially with revolutionary regimes (France in 1789; Italy in 1922; Cambodia in 1976), universal time remains fixed according to the Gregorian calendar, established in Papal Rome during the sixteenth century and globalized during the twentieth. In this system, Year Zero is Christ's attributed birthdate, and thus, for a self-proclaimed Universal Church, year 2000 is the most important *Anno Santo* for a thousand years.

Both London and Rome thus lay distinct and particular claims to the role of universal *axis mundi*,[7] points within global space which perform unique symbolic functions in articulating and radiating meaning within and beyond their ritual sites.[8] Vertical axiality, striking from the depths of space to the heavens, fixes centrality in both global space and time, as the French landscape architect Bernard Lassus (1998) argues. Once complete, the exploration of the formerly immeasurable in the horizontal direction is succeeded by an approach to the immeasurable verticals: the conquest of space, the depths of the sea and the earth. The surface we tread on is also a depth – immeasurable, vertical, obscure, as Lassus seeks to illustrate in his sketch of the 'vertical garden' proposed for the Parc de la Villette in Paris (figure 8.1). In the case of both London and Rome, therefore, the focus of millennial celebration is a domed structure, constructed at a strategic urban location: in Rome the point of St. Peter's martyrdom, foundational 'rock' for constructing a universal Church on Earth; and in London, the point where the 0^0 meridian meets the River Thames, artery of Britain's nineteenth-century global commercial empire.

While focused on these domed structures, the performative sites of the millennium in both cities connect beyond them into broader urban and global spaces. At the first Catholic Jubilee, proclaimed by Pope Bonifacio VIII in the year 1300 (Tripodi 1997),[9] pilgrims gathered from across Christendom to follow processional routes describing the form of the cross etched over Rome's urban space by the four apostolic Basilicas – St. Peter's, San Giovanni in Laterano, Santa Maria Maggiore, and San Paolo Fuori le Mura – in order to receive the indulgence offered in Bonifacio VIII's announcement (figure 8.2; cf. Norberg-Schultz 1980: 150–1).

More than two million pilgrims visited Rome in that first Jubilee year, at a time when its resident population amounted to no more than 50,000 (Morello 1996). The Hebraic tradition of Jubilee had originally mapped the identity, structure, and functioning of Jewish community in its place of residence. For imperial Catholicism, by contrast, Jubilee meant, above all, pilgrimage (Ciabattoni 1997), a sacred gesture enacted

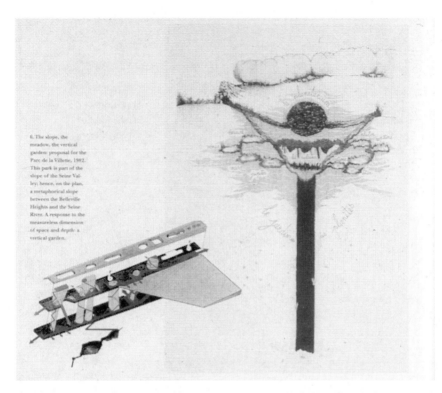

Figure 8.1 'The slope, the meadow, the vertical garden: proposal for the Parc de la Villette, 1982' (Lassus 1998). Reproduced by kind permission of Bernard Lassus

by movement to the center of a universal Church, with the processional geometry of the Christian cross inscribed into the fabric of the city. The year 2000 sees many millions of pilgrims from every country in the world processing formally or informally through Rome. The Jubilee 2000 is the first to be celebrated since the fall of the Berlin Wall, and millions of visitors are able to visit Rome from the territories of the former Soviet empire. The Catholic Jubilee is designed to reinvest the millennial moment with an explicitly sacred dimension by mapping a living, universal Church across the urban spaces of its global center. The four great basilicas – Christianized reminders of Augustus's empire *ad termini orbis terrarum* – and the routes that connect them, are intended to provide the focus of the Church's heavily funded renovation efforts for the Jubilee 2000. The municipal authorities, responsible for the city's

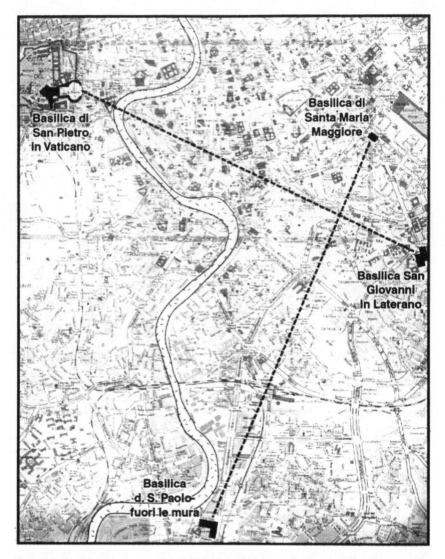

Figure 8.2 Sketch map of Rome showing the location of the four basilicas and the Christian cross which the route of the Jubilee pilgrims describes

transport and accommodation infrastructures, have the task of locating and transporting this vast temporary population within urban space. One architect has proposed an urban reconciliation in Rome between the world's two globalizing faiths by means of a new bridge across the Tiber at the junction of two urban axes, extending to St. Peter's and to a new Mosque respectively (Berdnadsky 1997). Sacred performances of the figure of the cross which the pilgrims' route describes are broadcast to the world by means of the most advanced communication networks, through radio, television, and the Internet, reaffirming the papal embrace of modern space/time-conquering technologies that dates back at least to Guglielmo Marconi's establishment of the Jesuit-run Vatican Radio in 1931, the first international broadcasting institution (Sacconi ca. 1997).

In London, the focus of the celebration of the millennium was located at an exhibition site close to the Royal Observatory in Greenwich. The Dome itself, situated on a formerly derelict peninsula jutting into the Thames, was opened with a spectacular firework display which stretched along the Thames, broadcast like the events in Egypt, Moscow, Paris, and New York in a continuous show on global television networks. Although planners claim that the Millennium Experience which occupies its space "must be durable and able to last 'forever'" ("Millennium Exhibition Saved...," 1997), the apparently flimsy materials and light structural elements that make up the stretched external membrane, with its flying saucer shape, suggest more the ephemeral aspects of time than the durability of eternity signalled by more conventional monumental materials and iconographies such as stone or Classical structural and decorative elements (figure 8.3). The Dome, in short, is "an installation, made to disappear" (Sinclair 1997: 10). Yet it also lies directly adjacent to other designed spaces intended to signify sustainability, most notably the 'Millennium Village,' a "high-tech, environmentally friendly regeneration" of the locally degraded urban social environment (Rowe 1997), whose naming makes an explicit gesture towards the local-global discourse of environmentalism – the late twentieth-century's most powerful secular religion. As Neil Smith (1998: 272) points out, "compared with the late 1960s and 1970s when the politics of nature erupted, *fin de millennium angst* about nature is widespread but of low intensity; we're all environmentalists now." The spiritual emphasis in London's millennial celebrations, while different from that of Rome, is equally intense. In his promotional speech for the Dome delivered at the 'People's Palace' of London's Royal Festival Hall in February 1998, the British Prime Minister spoke of the inspirational intent of the project (Blair 1998: 4): "I want today's children to take from it an experience so powerful and memories so strong that it gives them that abiding sense of purpose and unity that stays with them through the rest of their lives." Actual attendance figures have not quite lived up to such grandiose visions.

As in Rome, moreover, the millennial celebrations in London are designed to map broader patterns across urban space, in this case fol-

Figure 8.3   Millennium Dome from the Royal Observatory in Greenwich, London, June 1999 (photograph by Luciana Martins)

lowing the line of the Thames in an attempt to signal the revival of the river's traditional significance as London's lifeline, a function lost with the end of empire and the decline of the river's maritime transport role (figure 8.4). These initiatives along the Thames include both private schemes and those financed through the national Millennium Fund. A 443-foot diameter 'Millennial Wheel' – officially called the 'British Airways London Eye' – located opposite the Palace of Westminster in the traditionally transgressive entertainment space of the South Bank, carries visitors in 32 enclosed passenger capsules for a spectacular flight over the heart of the capital, "a demonstration of the use of renewable energy: solar cells will be built into the passenger capsules and will help power the ventilation, lighting and communication systems" (Murray and Stevens 1998: 102). While the Dome displays inward visions of Britishness, the more successful Wheel invites visitors to share London's outward horizons. A 'Millennial Footbridge,' the city's first new river crossing in a century, now connects the financial heart of London and St. Paul's Cathedral (the British capital's architectural response to Rome's basilica) to a new Tate Modern opened in May 2000 at Bank-

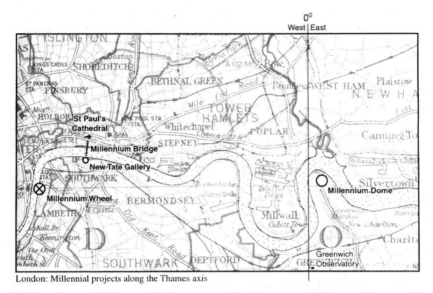

London: Millennial projects along the Thames axis

Figure 8.4 Sketch map of the Thames at London showing location of major Millennium projects

side. Free from motorized traffic, the Footbridge is a conscious attempt to revive the idea of public space in the modern metropolis. While locally significant in terms of urban regeneration, such projects as these are also explicitly designed to register London's claims to global centrality: once represented through the figure of empire (Driver and Gilbert 1999), this centrality is exercised today largely through mastery of financial space–time in the City. To quote the British Prime Minister again (Blair 1998: 1): "The clock strikes midnight on December 31st 1999. The eyes of the world turn to the spot where the new Millennium begins – the Meridian Line at Greenwich. This is Britain's opportunity to greet the world with a celebration that is so bold, so beautiful, so inspiring that it embodies at once the spirit of confidence and adventure in Britain and the spirit of the future in the world." Meanwhile, the BBC (which quickly followed Vatican Radio into the business of global broadcasting) offers the 'TV show of the century': the 'arrival of the new millennium' brought home to London "live from every corner of the UK and the globe."[10]

In the case of both London and Rome, then, we are presented with a symbolic mapping of the millennium based, not on territories, but on networks: both actual (in the buildings, churches, and processional pathways) and virtual (in their axial connections and linkages through modern technologies). Furthermore, the spatiality of millennium is more than conventionally monumental: it relies upon performances. Interactive exhibits filling the London Dome's central space develop a variety of themes: "where we live – local, national and global"; who we are – "body, mind and soul"; "what we do – work, rest, and play" (MacAsill 1997: 10). Of the Dome's various 'zones,' the most controversial in planning were the 'Body Zone,' the 'Faith Zone,' and the 'National Identity Zone.' In the first of these zones, visitors wander inside huge plastic models of human bodies by means of snaking tunnels, in order to view the internal spaces of their own corporeality. The Faith Zone was the cause of considerable controversy over the beliefs to be celebrated in a multifaith nation which still has an established Church. There was a marked reluctance on the part of traditional Christian religions, most notably the Anglican Church itself, to finance the project which was eventually underwritten by a Hindu business entrepreneur. In the National Identity Zone, that challenge was intensified by the erosion of a single narrative of national identity within Britain: the millennial year was also the first year of functioning of a new Scottish parliament and a Welsh assembly, devolving power to the constituent nations of the United Kingdom, while many former icons of nationhood are widely regarded as too narrowly sectional to embrace an open, liberal, and multiethnic society at the close of the twentieth century.

Such dilemmas indicate the impossibility of reaching a consensus on the Dome's own principal 'narrative': should it be sacred or secular, national or universal, historicist or futuristic, populist or didactic? Together with public debates over issues of surface/depth and transience/permanence raised by the architectural forms and moral significance of the Dome's contents, these dilemmas expose inherent contradictions in the celebration of a univocal, implicit universalist, discourse of Western millennialism within the polyvocal, localized, and multicultural society that Britain – and especially London – has become in the opening years of the twenty-first century. They may also reflect the condition of postmodernity, insofar as they expose the increasing fragility of a unifying discourse of 'nation' or 'people' and the dissolving integrity of a univocal and universal, but actually largely Western, discourse of human solidarity and rights. As Homi Bhabha (1994: 171) points out, from a postcolonial position, 'to reconstitute the discourse of cultural difference demands not simply a change of cultural contents and symbols; a replacement within the same time-frame of representation is never adequate.' In the case of the millennium, the geographical frame, as well as the historical, has to be negotiated. (In Rome, this process is perhaps rendered less problematic by the disciplines of

Catholicism – although, in reality, these too are highly contested – and their performative coordination through the Mass and similar sacred rituals.) The hegemonic choreographing of diverse peoples, either in Roman religious performance or in the theme-park festivities of the London Dome, cannot simply reactivate the ideals of a Western modernity that once symbolically encompassed a colonized globe. Nonetheless, it is impossible to ignore the fact that the physical spaces in which all these events are staged are already shaped by deposits of the past; the ghosts that haunt these locations are constantly being evoked, animating their *genius loci* (cf. Bell 1997; see Loukaki 1997 for similar dilemmas in Athens). The experience of Paris's bicentennial celebration in 1989 (*Magiciens de la terre*), which attempted a synthesis of both postcolonial polyvocality and claims to the universality of French revolutionary ideals in the streets of Haussmann's imperial metropolis, points to the tensions and contradictions involved in such mappings.[11]

Millennial Mapping and Performance

In January 1999, an Anglo-German team of astronomical cartographers announced the production of a new three-dimensional map of the universe: 'the largest ever map of the cosmos, . . . the definitive work' representing the state of scientific exploration of the space–time universe at the moment of the millennium (figure 8.5; Millar 1999). In the light of a 'decentered' contemporary condition, a notable feature of this millennial scientific image of the universe is that Earth itself is mapped into exactly the same position of cosmic centrality attributed to it in Classical Antiquity: the Archimedian point whence all the measurements have been calculated. If we abstract the geometrical structure of this map from the superclusters and voids it illustrates, what remains is the reproduction of a remarkably enduring Western cosmographic image, the hegemonic geometrical construction of sphere and cube. The cosmographic knowledge here represented is publicized as the product of 'disinterested' science, a presentation which, in the light of the past decade of critical reflection on both 'science' and 'cartography,' seems naive.

While few in the human sciences remain convinced by conventional claims of cartographic mimesis, they are more than ever fascinated by the power and poetic potentials of mapping as a mode of representation and communication of spatial relations. The reasons for the contemporary revisioning of maps and mapping are not difficult to detect. At a technological level, economic change, driven by advances in infor-

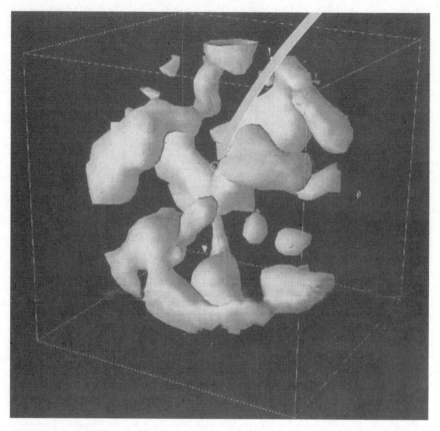

Figure 8.5   3-D map of the cosmos produced by an Anglo-German cartographic team

mation processing and by new and highly flexible financial and industrial production systems, has reworked the experience, meanings, and measures of space, rendering boundaries of all kinds permeable (Lévy 1991). Indeed, the concept and practice of precise and categorical separation and contiguous spatial 'fixing' inherent in boundary definition and conventional mapping (whose *sine qua non* is the bordering frame) represent an urge towards classification, order, control, and purification. From a postmodern perspective, these are defining features of a suspect modernity whose historical duration has been relatively brief, whose goals have always been compromised, and whose intellectual hegemony is today profoundly insecure. It is the spatialities of connectivity, network linkage, marginality, and liminality, and the transgression, permeability, and erasure of linear boundaries and hermetic categories by spatial 'flows' that are increasingly said to characterize experience in the late twentieth-century world. Such spatialities stimulate new forms of cartographic representation, for example, in the virtual spaces of the Internet, which show scant regard for the conventions of scientific mapping. These express not only the qualitative demands of new spatial structures and relations but also the altered disciplinary divisions and hierarchies they generate. In a world of radically unstable and mobile spaces and structures, it is hardly surprising that the idea of mapping should require rethinking. In this context, the humanistic cartographic image of a geocentric universal space offered to us by universalizing science is at once conservative rather than progressive, unconscious heir to a long heritage of cosmographic images.

The characteristically relational and constructed spatialities of a postmodern, non-essentialist geography offer both challenges and opportunities for mapping as a practice of knowledge-making and knowledge-communicating. Fixed and universal spatial coordinates, and the demands of scalar accuracy, iconic consistency, and naturalism in topographic representation are relevant to specific goals and contexts. Where such conventions are adopted, we need to be conscious of their implications and limitations. If such "things are not given, [but] are products of processes in particular times and places" (Massey 1999: 18), then the map itself becomes a discursive expression of an active and participatory geographical imagination, and thus liberated to participate in the creative processes of place-making. At the same time, we may expand our conception of the act of mapping beyond the conventional practices of surveying, measuring, sketching, and designing that have encompassed cartography in the West since the fifteenth century. Thus, for example, the British painter Stephen Farthing plans to put onto the Internet the latest of his series of topographical pictures, *The Knowledge*, which map his own itineraries and experience in such cities as Kyoto, Liverpool, and Buenos Aires. The image selected for London represents urban topography from a point south of the Thames near Waterloo (figure 8.6). Local people – indeed anyone – will be invited to participate in the artwork by adding to the virtual image, mapping their own memory, experience, or desire across its topography and downloading the resulting artwork into hard copy.

Mapping activities have, of course, never been entirely constrained by the demands of narrow mimesis; they have long opened spaces for memory and desire. Maps played significant projective and proactive roles in the projects of modernity: in the planning and design of cities, rural communities, and whole territories (postrevolutionary and Napoleonic France, Jeffersonian America, and the planned Brazilian

Figure 8.6 Stephen Farthing, *The Knowledge, SE1 West*, acrylic on canvas, 275 × 173 cm, 1996; reproduced by kind permission of the artist

capital of Brasilia offer the most obvious examples), in the appropriation of colonized lands, and in the construction of military strategy (Söderström 1996; Cosgrove 1999). Geographical and psychological studies of cognitive mapping in the 1960s and 1970s, as well as Situationist psychogeography, opened up the idea of mapping, beyond the technical activities of cartographers, by examining both the processes of 'mental mapping' in the minds of individuals and groups as they sought to negotiate material space, and the design role played by mapping processes in site planning (Lynch 1962; Gould and White 1974; Wood 1992). Performative mapping arises from the recognition that all human spatial activities, and not only cognitive ones, may be regarded as incorporating mapping acts and outcomes. Yet the map has a concreteness in its connection to material spaces. It is through this connection that the concept of performative mapping can move us beyond the purely constructivist position in understanding spatialities, wherein human relations with the material world become entirely discursive. It allows us to approach what Nigel Thrift (1999: 302, 308), following Heidegger, has labeled a 'dwelling' perspective, "anchored in an irreducible ontology in which the world is made up of billions of happy and unhappy encounters, encounters which describe a 'mindful connected physicalism' consisting of multitudinous paths which intersect."

From such a perspective, both the physical construction of celebratory spaces in London and Rome to mark the millennium and the various movements, encounters, and placements that occurred in these cities in the course of the year 2000, make up complex mappings, translocations, and inscriptions of spatiality in different locations and at different scales. In some respects, this is a banal observation. It is rendered more significant by the connection between millennial activities in London and Rome and the conventional time–space anchors adopted by Western mappings. Performances of the millennium in London and Rome are consciously and immediately controlled by the calendar, they mark a determined and determinate moment which, like a birthday or anniversary in private life, is not allowed to pass unmarked. There is no claim to permanence, the choreography is intended to designate both event and place as ritually appropriated: to gain the Plenary Indulgence the pilgrim must pass through the portals of St. Peter's at this particular time, any other time or place renders the gesture ineffective. In Merleau-Ponty's words, "the visible landscape before my eyes is not external to, but linked sympathetically with . . . other moments in time and in the past, but these moments are genuinely behind the landscape and simultaneously within it, not they and it side by side 'in' time" (quoted in Lassus 1998: 116). Countless individual pilgrimages to a common site and collective participation in ritual movements within that place – Rome or London – activate and inscribe its meaningful global location. Gathering also generates the fourth, temporal dimension of performative mapping, by activating *genius loci*, what Thrift (1999: 316–17) calls "the ecology of place . . . a rich and varied *spectral gathering*, an articulation of presence as 'the tangled exchange of noisy silences and seething absences,'" which emerges from the coincidence of imagination and action, memory, and desire. The "meaning of a place, its *genius loci*, depends upon the geniuses we locate there," and these ghosts and geniuses "constitute the specificity of historical sites" (Bell 1997: 837), perhaps nowhere more powerfully than in Rome and London.

One element in the complex, multiple affiliations that bind us to the visible and tangible world, performative mapping is a creative intervention in making rather than representing space, generating knowledge and insight: "the art of observing is combined with the art of inventing" (de Certeau 1985: 17). Like all creative interventions, however, it takes seriously the deposits of past experiences, the accretions which shape the physicality of places, whose intensity of presence varies in both linear and cyclical time or, more precisely, in spiral time. In Marcel Mauss's words (quoted in Lassus 1998: 140): "souls are mixed with things, things are mixed with souls," so that "through ghosts, we re-encounter the aura of social life in the aura of place." And if the morphology of the places thus constituted is discontinuous and fragmented – constellations or "lattices, archipelagos, hollow rings or patchworks" (Lewis and Wigen 1997: 200), formed and reformed according to the ghosts inhabiting and activated within them – so moments in time have a similar flexibility and arbitrariness, attaining significance and attachment through their inhabiting spirits, the performances that reanimate those spirits, and the countless but choreographed mappings that register those epiphanies. Thus the multiple positionalities, polyphonies, and other relativities so precisely delimited within the postmodern condition cannot avoid engagement with the deposits already mapped into the physical spaces of London and Rome, universalistic, imperial, and eurocentric though these might, in large measure, be. It is precisely in this metaphorical engagement that performative mapping finds its creativity.

Performative Mapping, Creative Arts, and Geographical Praxis

The concept of performative mapping thus reminds us of the significance of geography's dialogue with the creative arts. Its aim is to introduce (in fact to reintroduce, if we think of the long connections between mapping and imaginative art) a 'demeasurable' scale, a 'vertical' dimension that takes into account the 'immeasurable of the imagination,' to borrow Lassus's words. It is an opening for the enactment of desires, in the sense that performative mapping joins what is already there with what *could be* (De Certeau 1985: 17). Following Lassus (1998: 65–6), to map means to engage in a poetic archaeology of discovering 'scraps,' to enhance the heterogeneity of places. A powerful example of such creative use of past mappings is found in the design for the Imperial War Museum North in Manchester by the architect Daniel Libeskind, whose project takes 'shards' of the globe and places these parts of the shell randomly to connect the form of the earth's curvature and the idea of its 24-hour revolution to the issues of historical conflict being dealt with in the museum (figure 8.7). As Lassus (1998: 140) claims, "meaning cannot be built up again from the outside. It is deep down in the cultural entity that the recuperative movement must take place."

Other millennial reworkings of the globe may be found in Rome and London. Thus the artist Pat Naldi, while in residence at the British School in Rome (an institution housed in Edwin Lutyens' neoclassical pavilion for Rome's 1911 Universal Exhibition, and itself modeled on the façade of St. Paul's in London), engaged in activating various ghosts

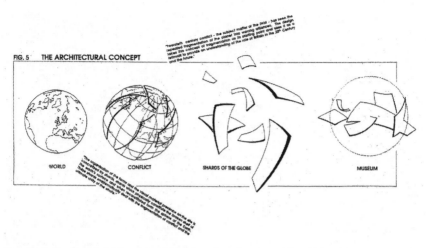

Figure 8.7   Daniel Libeskind, Design for the Northern War Museum, Manchester, UK

that haunt and connect her personal geography, and those of Rome and London, to universal space and time through a series of mapping projects. Her projects search out and transform various threads linking the British School building itself, the great collection of Ptolemy's *Geographia* (the West's foundational cartographic text, held at the Vatican to secure papal claims to universal authority through possession of the cartographic means to map the globe), the British claim to global imperialism, and the enduring Western privileging of visual knowledge. One of Naldi's installations, entitled *Neutral Zones*, concerns apparently 'blank' spaces on the map located between international frontiers. At either end of the atrium in the British School, two outlines in adhesive black plastic, unnamed but marked by coordinates of latitude and longitude, represent the neutral spaces between the British colony of Gibraltar and mainland Spain, and the Spanish colony of Ceuta and mainland Morocco respectively (figure 8.8). In Classical times, this was the location of the Pillars of Hercules, the edges of the mapped ecumene, and the spatial limits consciously broken by Iberian navigation in the opening years of Europe's modernist imperial adventure. Naldi's installations thus signal and locate acts of mapping – creative, often anxious, moments in connecting with the world, in which the map performs multiple functions: at once an animation of past engagement with space, a spatial embodiment of present knowledge, and a stimulus to further cognition. The artist's work reminds us that

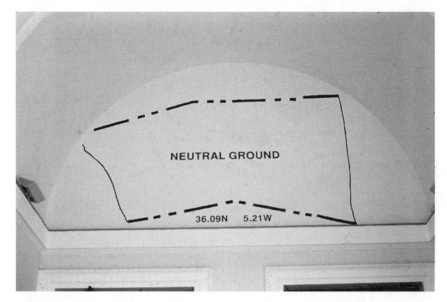

Figure 8.8   Pat Naldi, *Neutral Zones*, one side of a double slide artwork, projected onto the atrium of the British School at Rome, 1999; reproduced by kind permission of the artist

mapping in a flexible era has become a creative and critical intervention within broader discourses of space and the ways it may be inhabited, in which the imaginative and performative potential of cartography can be more actively mobilized.

Similar moves may be found in the world of advertising, postmodern society's most characteristic artform. Corporations draw upon stocks of standard advertising images to promote the idea of 'being global' much more than they specify the ways in which the companies concerned actually operate across the world (Thompson 1999: 28). Conventional cartographic images of the globe are deemed too specialized and insufficiently pleasing to succeed in expressing the abstract values required in corporate advertising. Thus mapping images do not have to be accurate or even realistic, and spatial expressions can be as inventive as possible, showing impossible angles, views, colors, resolutions, and situations. As Pickles (1992: 211) has noted, "the globe and the map have become so successful as symbolic images that their 'shadows' can be presumed in images which contain no map form at all," relying for meaning on preformed mental stocks of global and topographic imagery to trigger messages and associations. The images perform for a visually literate audience purely as gestures towards material space.

Our use of the term 'performative mapping' to describe the array of material structures, and individual and social activities that constitute the millennial geographics of Rome and London reflects a developing creativity and imaginative freedom within formal geographical study which has embraced the insights but moved beyond the rhetorical radicalism of Situationist claims about such practices. It is significantly removed from belief in an essential 'Geography,' circumscribed in epistemology, theory, content, or methods. While human fascination with spatial difference and the difference that space makes in the world certainly endures, it is variously and variably articulated and represented, both historically and culturally. The first part of the twentieth century was dominated by a Geography concerned to describe and amplify the connections between territory and power in an era of Western nationalism and global imperialism; the century's middle years saw Geography drawn into and shaped by the contest for ideological hegemony between two grand narratives of sociospatial order and human liberation, each aligned to a universalizing discourse of science. Millennial Geography today is being shaped in the context of more than a decade of post-cold war geopolitics, post-Fordist economics, globalized informatics, and postmodern philosophy. In this cultural context of flexible identity, complexity, and destabilized meaning, there is an ever-increasing emphasis on creativity as much as on rational cognition in our spatial practices, related to the current awareness that "knowledge is never distant either from our deep emotions or our aesthetics" (Dening 1996: 217). The millennium also opens with the decisive collapse of distinct genres and hierarchies in the realm of what was traditionally regarded as Art, releasing into daily life those activities conventionally closed within the bounds of an 'art world.' Geography can no more avoid engagement with these cultural shifts than it could with imperialism, Fordist production, or scientism.

The idea of performative mapping connects to an area of geographical praxis that is currently under intense exploration. In this sense, Stéphane Quoniam (1988) can be seen as a pioneer: his mappings of Arizona combined effectively the art of observing with the art of inventing, using multiple media of representation. In their digs into archives, scourings in museums, their fieldwork, and other performances, geographers are today engaging ever more directly with the creative arts. Catherine Nash's collaborations with the artist Kathy Prendergast provide a clear example. Prendergast's artwork *Lost* (1997), the first part of a project to create an *Atlas of Emotions*, relied on an archive of emotional toponyms and grid references in Canada derived from Nash's geographical research. According to Nash (1998: 5), Prendergast's artwork "harnesses the pleasures of geography's traditional empirical and encyclopedic impulses but also combines the grid-referencing accuracy of cartography with the unpredictability of emotion." It is this unpredictability of emotion – the 'ordinary inaccessibility and involuntariness' of bodily knowledge – that also drives Jeremy Foster's (1998: 341) research. John Buchan's *Hesperides* provided the material for Foster's study of how the landscape of the early twentieth-century Transvaal, as depicted by Buchan, appeared "to embody complex and highly abstract structures of feeling" (Foster 1998: 341). Similarly, in researching nineteenth-century British 'mappings' of 'tropicality' on the Brazilian littoral, Luciana Martins (1999; 2000) examined navigators' log-books, naturalists' diaries, and artists' sketchbooks in order to understand the role of imagination and memory in engagements with new landscapes. While the resulting images necessarily shared with the practices of painting, drawing, and surveying a set of codes integral to the visual culture of early nineteenth-century Europe, as well as being material evidence of Britain's global imperial project, they were always more than simply an 'expression' of these wider historical realities. Inscribed in the very materiality of practices and products are also traces of the travelers' performances. They are vestiges of memory, shaped through a constant process of negotiation between translocated images: brought by the traveler, found in the landscapes presented to eyes and bodies, tracked over, and variously contacted, remembered, and forgotten. Travelers also had to make themselves understood to others whose senses had passed through different routes, so that the maps of departure were not those of return (cf. Dening 1980; Carter 1987; Beer 1996). Representation is thus always a contingent and problematic process, a performance, rather than the *automatic* reflex of a 'way of seeing.' Artefacts produced 'in transit' bear the traces of such performances. Pieced together, the scraps of meanings, memories, and desires that haunt the images of particular locations – Arizona, Canada, Rio, the Transvaal, London, Rome – disclose some tracks of past and present performative mappings.

Conclusion

The inescapable ambivalence of these processes of geographical representation opens up questions for creative investigation, rather than closing them down. This is not in the least to suggest that creativity and artistic imagination have had no place in geography in the past. That creative imagination animates any scholarly project is a commonplace in postmodern revisionings of scientific method. Creative imagination makes present the spaces, landscapes, movements, and circulations in any geographical representation. But millennialism and the performance of the millennium offer the opportunity to signal the current opening of the discourse of geography towards the creative arts – visual, graphic, and performative. Of course the discipline of Geography has had a sustained historical engagement with the visual arts, especially through mapmaking, with the performative arts in theatrical and urban design, with poetry, romance, travel writing, and cinema, though such engagements have generally cast 'the Geographer' more in the role of critic than creator. Alexander von Humboldt memorably signaled the importance of an artistic sensibility for the practice of natural philosophy and for understanding the cosmos. Revived interest in his work suggests that he is among the ghosts that animate the current renewal of interest in the creative capacities of geographical discourse (Greppi 1996; Harvey 1998). But at the millennium, the cosmos itself is no longer a stable category within which old dualities of culture and nature, animate and inanimate, self and other, obtain. The space for creativity formerly occupied by the imaginative arts is thus enormously expanded, epistemologically and technically, and leaches into the humanities and sciences. At the millennium, we are beginning to rediscover the fact that the boundaries formerly separating the laboratory or the field from the studio, the screen, and the stage are distinctly permeable. As artists, architects, and film-makers draw upon and begin to contribute to geographical scholarship, geographers feel freer to engage directly in creative representation, curatorship, and the aesthetics of display. Millennial geographics extend far beyond the events of the year 2000 in London and Rome.

Acknowledgments

We would like to thank Felix Driver for his insights and comments; our debt extends to colleagues of the 'Landscape Surgery' group at Royal Holloway for their suggestions on a previous version of this chapter. We are also grateful to Alessandro Scafi for helping us through the paths of Vatican Radio, Ciro Fusco for providing information on the preparations of the Jubilee in Rome, and Pat Naldi and Stephen Farthing for allowing access to their work in progress. A version of this chapter appeared in the *Annals of the Association of the American Geographers* 90: 97–113.

NOTES

1   For studies of Rome and London as imperial cities, see Atkinson and Cosgrove (1998) and Driver and Gilbert (1998; 1999).
2   For a detailed historical account of the calendar as a scientific, cultural, religious, and social object, see Le Goff (1984). The recently resurrected Situationist group: The London Psychogeographical Association in one of its web pages called for people to 'say no to the millennium' in part by adopting the 'Modern Khemetic Calendar,' based on Egyptian worship (www.unpopular.demon.co.uk/lpa/elpan018/elpan018.html).
3   The latter is captured by John Donne in the first lines of his Holy Sonnet: 'At the round earth's imagined corners . . .'; on the Tower of the Winds at Athens, see Raff (1978–9) and Jacob (1992).
4   These dates are convenient markers, 1450 signifying the beginnings of European navigation and mapping along the Atlantic coasts, and 1950 being the year in which flights over Antarctica determined the precise outlines of the last continent to be mapped onto the globe.
5   Recent interest in the history of the fixing of longitude and the construction of global time/space is indicated by Umberto Eco's novel, *The Island of the Day Before* (1995), and Dava Sobel's, *Longitude* (1996). On French resistance to the adoption of Greenwich Mean Time (GMT), see Mattelart (1996: 163–4).
6   This estimate was made in February 1999 by the *Agenzia Romana per la Preparazione del Giubileo*, the body in charge of the establishment of the overall action plan, visitor services, and public information concerning the Jubilee.
7   *Axis mundi* refers to a place where cosmic time (*in illo tempore*) is fixed in terrestrial space, a site of profound symbolic and cultural significance. Conventionally, these are places of hierophany or epiphany. Such points (Delphos, Mecca, Beijing, Jerusalem) are marked by permanent spatial inscriptions and monuments (ceremonial routes, altars, columns) and elaborate ritual and ceremony (sacrifice, song, immolation). For a classical study of this theme, see Eliade (1959).
8   For a recent investigation of the symbolic centrality of cities in mythical spaces, see Sennett (1994).
9   Tripodi (1997) notes that Bonifacio VIII fixed the Jubilee to be celebrated every hundred years. The interval was reduced to fifty years by Pope Clemente V, in 1342, and to 33 years by Pope Martino V. Finally, in 1450, Pope Nicolo V reduced the interval between celebrations to 25 years, which is still current. To date, 25 such Jubilees and 96 extraordinary holy years (nominated to celebrate important events) have been celebrated.
10 BBC Millennium, special supplement of *Radio Times*.
11 For a critical account of *Magiciens de la terre*, see Buchloch and Martin (1989).

REFERENCES

Atkinson, D. and D. Cosgrove (1998). Urban Rhetoric and Embodied Identities: City, Nation and Empire at the Vittorio Emanuele II Monument in Rome, 1870–1945. *Annals of the Association of American Geographers* 88(1): 28–49.

Baudrillard, J. (1998). The End of the Millennium or The Countdown. *Theory, Culture & Society* 15(1): 1–9.

Beer, G. (1996). *Open Fields: Science in Cultural Encounter*. Oxford: Clarendon Press.

Bell, M. M. (1997). The Ghosts of Place. *Theory and Society* 26: 813–36.

Bennett, T. (1996). *The Birth of the Museum: History, Theory, Politics*. London and New York: Routledge.

Bhabha, H. K. (1994). *The Location of Culture*. London: Routledge.

Blair, T. (1998). Why the Dome is Good for Britain. Press release, Royal Festival Hall, 24 Feb.

Berdnadsky, C. (1998). Personal communication, June.

Buchloch, B. H. D. and J.-H. Martin (1989). Interview. *Third Text* 6: 19–27.

Carter, P. (1987). *The Road to Botany Bay: An Essay in Spatial History*. London and Boston: Faber and Faber.

Ciabattoni, A. (1997). Il Pellegrinaggio: Un secolare strumento di comunicazione tra le genti. *Tertium Millennium* 1(3): 77–80.

Cosgrove, D. (1994). Contested Global Visions: 'One-World, Whole-Earth,' and the Apollo Space Photographs. *Annals of the Association of American Geographers* 84(2): 270–94.

Cosgrove, D. (1999). Introduction: Mapping Meaning. In *Mappings*, ed. D. Cosgrove. London: Reaktion Books.

Cosgrove, D. (2000). *Apollo's Geography: Global Images and Meanings in Western Culture*. Baltimore: Johns Hopkins University Press.

Debord, G. (1955). Introduction to a Critique of Urban Geography. *Les Lèvres Nues* 6: www.slip.net/~knabb/SI/urbgeog.htm.

de Certeau, M. (1985). Pay Attention: To Make Art. In *The Lagoon Cycle: Helen Mayer Harrison/Newton Harrison*, ed. L. M. Colvin. Ithaca, New York: Herbert F. Johnson Museum of Art, Cornell University.

Dening, G. (1980). *Islands and Beaches: Discourse on a Silent Land. Marquesas 1774–1880*. Hawaii: University Press of Hawaii.

Dening, G. (1996). *Performances*. Chicago: University of Chicago Press.

Driver, F. and D. Gilbert (1998). Heart of Empire? Landscape, Space and Performance in Imperial London. *Environment and Planning D: Society and Space* 16: 11–28.

Driver, F and D. Gilbert, eds. (1999). *Imperial Cities: Landscape, Display and Identity*. Manchester: Manchester University Press.

Eagleton, T. (1999). Fictions Etched by Rain in Rock. *Times Higher*, Feb. 26: 27.

Eco, U. (1995). *The Island of the Day Before*. London: Secker and Warburg.

Eliade, M. (1959). *The Sacred and the Profane: The Nature of Religion.* New York: Harcourt Brace.

Fallows, J. (1999). The Y2K Specter: Hurry Up Please It's Time. *New York Review of Books,* 46: 29–34.

Foster, J. (1998). John Buchan 'Hesperides': Landscape Rhetoric and the Aesthetics of Bodily Experience on the South African Highveld, 1901–1903. *Ecumene* 5(3): 323–47.

Goodwin, B. (1997). Complexity, Creativity, and Society. *Soundings* 5: 111–22.

Gould, P. and R. White (1974). *Mental Maps.* Harmondsworth: Penguin.

Greenhalgh, P. (1988). *Ephemeral Vistas: A History of the Expositions Universelles, Great Exhibitions and World's Fairs, 1851–1939.* Manchester: Manchester University Press.

The Greenwich 2000 Network. (1997). Millennium Exhibition Saved by Prime Minister Tony Blair.

Greppi, C. (1996). *Intorno a Humboldt (nove interventi 1992/96).* Ferrara: Università degli Studi di Ferrara.

Harvey, D. (1998). The Humboldt Connection. *Annals of the Association of American Geographers* 88(4): 723–30.

Jacob, C. (1992). *L'Empire des cartes: approche théorique de la cartographie à travers l'histoire.* Paris: Albin Michel.

Lassus, B. (1998). *The Landscape Approach.* Philadelphia: University of Pennsylvania Press.

Le Goff, J. (1984). Calendário. In *Enciclopédia Einaudi, Vol. 1: Memória-História.* Porto: Imprensa Nacional-Casa da Moeda.

Lévy, J. (1991). A-t-on encore (vraiment) besoin du territoire? *Espaces et Temps* 51(2): 102–42.

Lewis, M. W. and K. E. Wigen (1997). *The Myth of Continents: A Critique of Metageography.* Berkeley: University of California Press.

Loukaki, A. (1997). Whose Genius Loci?: Contrasting Interpretations of the 'Sacred Rock of the Athenian Acropolis.' *Annals of the Association of the American Geographers* 87(2): 306–29.

Lynch, K. (1962). *Site Planning.* Cambridge, Mass.: MIT Press.

MacAskill, E. (1997). Innovative Architect Given Dome Role. *The Guardian,* 4 Nov. 10.

Martins, L. L. (1999). Mapping Tropical Waters: British Views and Visions of Rio de Janeiro. In *Mappings,* ed. D. Cosgrove. London: Reaktion Books.

Martins, L. L. (2000). *O Rio de Janeiro dos Viajantes: O Olhar Britânico, 1800–1850.* Rio de Janeiro: Jorge Zahar Editor.

Massey, D. (1999a). Issues and Debates. In *Human Geography Today,* eds. D. Massey, J. Allen, and P. Sarre. Cambridge: Polity Press.

Massey, D. (1999b). Spaces of Politics. In *Human Geography Today,* eds. D. Massey, J. Allen, and P. Sarre. Cambridge: Polity Press.

Mattelart, A. (1996). *The Invention of Communication.* Minneapolis and London: University of Minnesota Press.

Millar, S. (1999). This is the Largest Ever Map of the Universe and You Are Here. *The Guardian,* Jan. 30: 1.

Morello, G. (1996). Gli Anni Santi: Storia e Immagini. *Tertium Millennium* 1(1): 77–82.

Murray, P. and M. Stevens, eds. (1998). *New Urban Environments: British Architecture and its European Context*. Munich: Prestel.

Nash, C. (1998). Mapping Emotion. *Environment and Planning D: Society & Space* 16: 1–9.

Norberg-Schulz, C. (1980). *Genius Loci: Towards a Phenomenology of Architecture*. London: Academy Editions.

Pickles, J. (1992). Texts, Hermeneutics and Propaganda Maps. In *Writing Worlds: Discourse, Text and Metaphor in the Representation of Landscape*, eds. T. J. Barnes and J. S. Duncan. London: Routledge.

Pinder, D. (1996). Subverting Cartography: The Situationists and Maps of the City. *Environment and Planning A* 28: 405–27.

Pred, A. (1995). *Recognizing European Modernities: A Montage of the Present*. London and New York: Routledge.

Prendergast, K. (1997). Lost. *Environment and Planning D: Society & Space* 15: 663–78.

Quoniam, S. (1988). A Painter, Geographer of Arizona. *Environment and Planning D: Society & Space* 6: 3–14.

Raff, T. (1978–9). Die Ikonographie der mittelalterlichen Windpersonnifikationen. *Aachner Kunstblatter* 48: 71–318.

Rowe, M. (1997). The Future is Green in Prezzaville-on-Thames. *Independent*, 26 Oct.: 12.

Sacconi, E. V. (ca. 1997). *Radio Vaticana*. Vaticano: Tipografia Vaticana.

Sennett, R. (1994). *Flesh and Stone: The Body and the City in Western Civilization*. London and Boston: Faber and Faber.

Sinclair, I. (1997). Mandelson's Pleasure Dome. *London Review of Books* 19(19): 7–10.

Smith, N. (1998). Nature at the Millennium: Production and Re-enchantment. In *Remaking Reality: Nature at the Millennium*, eds. B. Braun and N. Castree. London and New York: Routledge.

Sobel, D. (1996). *Longitude*. London: Fourth Estate.

Söderström, O. (1996). Paper Cities: Visual Thinking in Urban Planning. *Ecumene* 3: 249–81.

Thompson, V. (1999). New Globes for Old: Uses of the 'Oxford Globe.' BA dissertation, Department of Geography, Royal Holloway, University of London.

Thrift, N. (1999). Steps to an Ecology of Place. In *Human Geography Today*, eds. D. Massey, J. Allen, and P. Sarre. Cambridge: Polity Press.

Tripodi, P. (1997). Rome Prepares for the Millennium. *Contemporary Review* 270 (1576).

Wood, D. (1992). *The Power of Maps*. London: Routledge.

9

Postmodern Temptations

Claudio Minca

Any search for identity, legitimate at the existential level, and at times
poetically fruitful, easily carries with it the mistaken deformation of the
social-historical real. The search for such an identity implies, more or less
consciously, an effort to grasp at an essence, at some more or less constant, permanent dimension within, inventing and accentuating analogies,
resemblances, rather than catching the transformations and distinctions
proper to all historical phenomena.

Claudio Magris, 1996

Introduction

The 'virtual' critical debate of the past two decades which has focused
its attentions upon the crisis of representation and the consequent need
for a reevaluation of the geographical models born of modernity has
opened the way towards the construction and legitimation of an 'other'
geographical praxis; a praxis capable of escaping from the closed spaces
framed by the logic of modernity.

Having dismantled the very foundations of positivist assumptions
that envisioned geographical representations as potentially faithful
mirrors of reality, what has emerged from this debate is the progressive
acceptance of the idea that we can identify a dialectical influence
between description and thing described; the by-now common assertion
according to which geographical representations not only narrate reality
but also contribute to its construction. The extent of this reciprocal
influence, however, remains, as yet, quite problematic.

A second point of contention brought out by the critical debate
among geographers has centered around the relationship between
society and the spaces that it inhabits, plans, and transforms. We have

moved away from understandings of social space as a mere spatial manifestation of the sociocultural dynamics which produce it – and towards more complex conceptualizations which, again, point to a reciprocal influence: thus, spaces both as the 'result,' as the expression of some social dialectic, but also spaces as a determining factor in the construction and constitution of the social dialectic itself. Again, however, a codification of this mutual influence remains problematic.

The consolidation of this position within geographical thought has resulted, above all, in a critique of the profound paradoxes and contradictions inherent in 'modern' ways of conceptualizing space and describing the world. This critical vision, however, has found itself, for the most part, hard pressed to overcome that exquisitely modern habit of reasoning in dualistic terms: that is, of envisioning the map and the territory as two rigorously and necessarily distinct objects; of considering society and space as two dimensions – associated, perhaps, but clearly distinguishable one from the other; of seeing the representations of the world and the world itself as two cleanly/clearly separate realms. The persistence of such a dualistic vision is evident, I would hazard, even within postmodern geography – an integral component of the critical forum that I alluded to above. I should note that with the 'postmodern' label I intend both the reflection on the postmodern approach to things geographical as well as the identification and analysis of so-called postmodern spaces, of spatialities which reflect a determined socio-economic 'condition,' perhaps even a 'postmodern epoch.' In this chapter, I hope to focus upon the possibilities – and the meanings – of a geographical praxis capable of transcending this latent dualism. I ground my reflections within the work of Timothy Mitchell (1988) and, in particular, his critique of the 'metaphysics of representation' that he develops so aptly in the pages of *Colonizing Egypt.*

What I will attempt to demonstrate is that the above dualism lies at the very heart of the mechanism of power instituted by modern logic in its attempts to colonize the Other – and, in so doing, the Self as well. A mechanism which, by separating the representations of power from the logic of power itself has, in some fashion, succeeded in objectifying and naturalizing this logic, thus constraining our analytical categories, our very ways of reasoning, into the closed spaces of this damning dualism.

The 'metaphysics of representation,' whose dynamics I shall attempt to delineate in the paragraphs to follow, obscure the workings of power through the logic of the 'world-as-exhibition': that is, through the construction of a separation between a material dimension, intended as the representation of some grand idea/ideal, of some 'external' meaning – and a conceptual, structural dimension, that is, meaning itself; always 'out there,' always representable, though never present, simply re-enacted within the materiality of the world. According to Mitchell (1988), it is this logic that has produced the 'regime of certainty' which has legitimated a distinct vision of the Other – and thus of the world as a whole. A regime of certainty which rests upon the construction of a set of pure, crystalline identities, defined and definable only as oppositions, as negations of the multiple alterities created by the representational order. It is this very logic which has managed to 'colonize' our relationship with space, with the world; which has inspired the practice of modern geographies; which has contributed to the 'creation' of the modern state by camouflaging the cartographic logic which underlies it; which has constructed a *homo geographicus* as one of a series of conceptual containers – 'culture,' 'society,' 'history,' 'the nation' etc. – with respect to which he can be individuated, isolated, and described. The map, the classic tool of geographical descriptions of the world and its various 'parts,' similarly embodies the logic of such a 'colonization.' It is through its simulation of more or less reliable reproductions of the territory, that mappings have succeeded in naturalizing the continual evocation of some underlying, objective reality which only awaits to be unveiled, to be narrated; a reality whose order – and whose very existence – necessarily depend upon its representability.

These are just some of the theoretical assumptions adopted within this chapter in order to isolate and reveal the power inherent in geographical metaphors as conceived within the logical 'cage' framed by the metaphysics of representation. My exploration intends to investigate the possibility of a geographical praxis 'other' than that imposed by the modern project: a praxis capable not only of denouncing and deconstructing the modern representational game, but also of transcending the closed spaces of the metaphysics of representation, proclaiming if not the demise, then at least the death throes of cartographic reason.

Behind and Beyond the World as Exhibition

The metaphysics of representation, denounced by Mitchell (1988) as that 'theological effect' which has facilitated the modern West's colonization of the Other (as well as of our own ways of conceptualizing the world) rests upon the idea of the world-as-exhibition: that particular theoretical apparatus which has allowed for the (material as well as conceptual) colonization of the world through the adoption and legitimation of a radical separation between reality and representation. A separation whose logic has succeeded in 'framing' the entirety of the

material and symbolic manifestations of the social within some structure, within a sort of external order; in other words, within a conceptual sphere capable of establishing what Mitchell (1988) terms a
'hierarchy of truth.' It is by virtue of this very hierarchy that modern
reason has produced a 'regime of certainty' associated with the functions and meanings vested within its representations: a 'regime of certainty' that has allowed for the construction of a dualistic vision of the
world; a vision capable of imposing its *own* order upon the world, as
though it were *the* natural order, an order inherent to all things, always
existing, and only now finally 'revealed.' It is by isolating the strategies
of the world-as-exhibition and demonstrating the logical mechanisms
which have allowed it to perpetuate the workings of power that
Mitchell's analysis begins to unravel this regime. Let us attempt to
briefly reconstruct his argument.

Mitchell's (1988) analysis begins with the increasing globalization of
capital in the second half of the 1800s and its architectural expression
within the principal European capitals, Paris above all. In particular,
Mitchell highlights the emergence of large department stores, galleries,
and shopping arcades, envisioned both as spaces for the exhibition of
goods, but also as new spaces of urban socialization; a tendency which
will culminate with the realization of the Universal Exhibitions which,
within their walls, shall attempt to reproduce spaces which recall both
the past and the triumph of the modern present, far-off lands as well as
the 'external' urban environment which surrounds the Exhibition itself.
This fashion for the reconstruction and recontextualization of goods
would meet with resounding success, becoming not only a popular
attraction for the thousands of individuals who visited the Exhibitions
but, perhaps even more importantly, instituting and reinforcing a whole
series of stereotypes of places and events in the European collective
imaginary; a series of stereotypes, of ideal types which, through spectacularization, would evoke "somehow, some larger truth" (Mitchell
1988: 6).

Within today's theme parks, shopping malls, holiday villages (as well
as their residential counterparts, that is, gated communities), revitalized
waterfronts, and rediscovered urban 'historical districts' and downtowns we find an exaggerated reproduction of this practice of a
thematic (and often hyperreal) recontextualization; a practice which,
though certainly differing in the techniques adopted, perpetuates the
very same emphasis on the spectacularization of geographical and historical representations. (We should also note here that it is precisely
these fragments of the contemporary urban context which have come
to symbolize the notion of postmodern spaces within a large part
of geographical analyses on the topic.) The emphasis placed upon the

representation's faithful reproduction of the original (that is, the 'external' environment), an emphasis similarly visible within our present-day exhibitions and Disney-esque spaces, could lead us to believe that the goal of the exhibition is that of best representing that 'external' original and, therefore, that the 'quality' of a representation lies precisely in its ability to provide the closest possible approximation of the thing represented – as, following the Cartesian logic, every 'proper' model should. Following the above line of reason, we could even conclude that in some cases such reproductions might attempt to simulate reality in order to replace/supplant it, dissolving its meaning through its very representation. Yet such reasoning would find us, again, prisoners of the metaphysics of representation; of that particular logic which allows us to reflect upon the quality of representations with respect to some 'reality' – but which never questions the representability of such a 'reality' itself. To escape from the cages of modern reason we must, therefore, seek out a different path.

According to Mitchell, the emphasis on and the spectacularization of the representational 'moment' within the exhibitional spaces, sanctions in an apparently clear and definitive manner the existence of a divide, of a clear separation, between the representation and 'external' reality, creating a sort of illusion in which all that which is 'inside' the exhibitional space is pure representation, while all that remains 'outside' is, in some way, reality. And this reality is . . . 'real' precisely in its representability as such. How was this representational effect created in the exhibitions of the late 1800s? First of all, the external world was represented as though on stage, as a series of visible, classifiable objects; observable and observed, however, only from a determined position, as it was "the distance that was the source of their objectness" (Mitchell 1988: 11). The certainty of the existence of an external reality was reinforced by three fundamental principles upon which every exhibition was founded:

1 *The apparent realism of the exhibits.* A realism which, it is important to note, was not intended as some veiled simulation of reality, but rather to create quite the opposite effect: the accuracy of the representation and its extraordinary power in evoking some external reality only reinforced the very idea of the *existence* of such an external reality: "It was precisely this accuracy of detail that creates certainty, the effect of a determined correspondence between model and reality" (1988: 7).
2 *The organization of the representations around a common center.* As Mitchell (1988: 8) notes, "the clearly determined relationship between model and reality was strengthened by their sharing of a

common center." At the Paris Universal Exhibition of 1889, "a model of the panorama of the city stood at the center of the exhibition grounds, which were themselves laid out in the center of the real city. The city in turn presented itself as the imperial capital of the world, and the exhibition at its center laid out the exhibits of the world's empires and nations accordingly . . . The common center shared by the exhibition, the city and the world reinforced the relationship between representation and reality, just as the relationship enabled one to determine such a center in the first place" (1988: 8).

3 *The position of the visitor as the occupant of this central point.* "What distinguished the realism of the model from the reality it claimed to represent was that this central point had an occupant, the figure on the viewing platform. The representation of reality was always an exhibit set up for an observer in its midst, an observing gaze surrounded and set apart by the exhibition's careful order. If the dazzling displays of the exhibition could evoke some larger historical and political reality, it was because they were arranged to demand this isolated gaze. The more the exhibit drew in and encircled the visitor, the more the gaze was set apart from it, as the mind is set apart from the material world it observes" (1988: 9).

The apparent realism of these reproductions[1] created a very particular relationship between the object-representation (which appeared to lay within arm's reach) and the visitor. Thus, "to the observing eye, surrounded by the display but distinguished from it by the status of the visitor," the representation "remained a mere representation, the picture of some strange (external) reality" (1988: 9). Even more importantly, this particular relationship operated within a double distinction to which I shall return time and time again within the pages of this chapter: the distinction between the visitor and the exhibit, and the one between the exhibit and that which it represented. In other words, the representation was set apart from the 'reality' which it claimed to represent, just "as the observing mind was set apart from what it observed" (1988: 9) – that is, the representation itself.

Let us, then, attempt to unravel the logic of the world-as-exhibition. The acceptance of the separation between the inside of the exhibition (the representation) and the external world (reality) was certainly favored by the clear spatial segregation of these realms: a segregation which, however, created the expectation of 'finding' the 'real world' outside, once having left the gates of the exhibit. Yet, more often than not, "the real world beyond the gates turned out to be rather like an extension of the exhibition" (1988: 10), for the 'external' modern urban spaces were themselves constructed and interpreted as representations of some other 'reality'; at times even as the spectacularization, the reflection, the materialization of some (external) will, plan; the expression of a social whole and its complex of power relations. Yet if such 'extended exhibitions' did, in fact, transcend the confines of the exhibitional space, could we not conclude then (taking this line of reasoning to its extreme) that, at the end, all of the (external) world is nothing more than a series of representations – and that which we consider 'reality' remains, always and everywhere, unreachable, due to the filter imposed upon it by our linguistic and conceptual categories?[2] In this conceptualization, no significant difference exists between the interior realm of the exhibition and the external world of the 'real' city; both act as symbolic spaces; both are representative of something which could also be but another representation in an infinite series of citations. Such a reading, however, would imply the acceptance of a discursive circularity which would take us away from the investigation of the workings of power which we intend to undertake.

An alternative critique could envision the exhibitional spaces as a mystification, since intentionally conceived to camouflage the power relations which uphold the relations of production, presenting 'distorted' images of the world in order to disguise the exploitation and alienation inherent in the capitalist order.[3] Yet such an understanding, similarly, remains ensnared within the metaphysics of representation: assailing the exhibition's representations as artificial and sinister misrepresentations of a reality which is entirely more complex from a sociopolitical standpoint. The materialist critique thus simply sanctions the very same divide between representation and reality theorized by the logic of the world-as-exhibition, merely contesting the latter's interpretative choices.

Mitchell, in fact, stresses that neither of the above two critical positions consents to an isolation of the 'theological effect' implicit within the metaphysics of representation since both, rather, merely reproduce the assumption of a distinction between some interior (the representation) and some exterior (reality) realm; as both position a subject who observes the world in a detached fashion, thus creating the appearance of order: "an order that works by appearance." As Mitchell (1988: 60; emphasis in original) notes:

> The world is set up before an observing subject as though it were the picture of something. Its order occurs as the relationship between observer and picture, appearing and experienced in terms of the relationship between the picture and the plan or meaning it represents. It follows that the appearance of order is at the same time an order of appearance, a hierarchy. The world appears to the observer as a relationship between

picture and reality, the one present but secondary, a mere representation, the other only represented, but prior, more original, more real. This order of appearance is what might be called the *hierarchy of truth*.

Just as in the case of the modern Orientalist construction of the Other, it is within this hierarchy that *all* of reality is recognized and legitimated as such; it is within the hierarchy of truth that reality is ordered through the system of representations adopted to narrate it:

> Believing in an 'outside' world, beyond the exhibition, beyond all process of representation, as a realm inert and disenchanted – the great signified, the referent, the empty, changeless Orient – the modern individual is under a new and more subtle enchantment. The inert objectness of this world is an effect of its ordering, of its setting up as though it were an exhibition, a set-up which makes there appear to exist apart from such 'external reality' a transcendental entity called culture, a code or text or cognitive map by whose mysterious existence 'the world' is lent its 'significance'. (1988: 62)

What emerges is a specific 'method' for the ordering, the distribution, the framing of reality – and it is this method which creates the seemingly insurmountable divide between an internal, representational dimension – and an external, conceptual, ordered reality. And it is precisely within this logic that the modern individual, according to Mitchell, has learned to experience reality: that is, through an ordered system of representations, eternally destined to recall some larger truth, some broader plan, some nonmaterial dimension whose materialization is but a reenactment. A world, in other words, seen and perceived as an exhibition of reality; as an infinite representation of something which stands behind and beyond (it), which grants meaning to that which we see/perceive, to some symbolic realm which is infinitely representable – but never truly present. The extraordinary power of the metaphysics of representation functions precisely by virtue of this non-accessibility of reality – and the concurrent faith in its representability. Corroborated by the Cartesian vision of a detached, observing Mind which classifies external reality within the order of its own models, the modern logic of representation thus frames/cages our 'access' to the world; our ability to relate to it, to its representability and, at base, to the order proposed by these representations as some interior 'mirror' of an external reality. Therefore, *it is not so much the order of representation that is put into question here but, rather, the very relationship between reality and its representability*: a relationship that becomes a formidable tool in veiling the mechanisms which regulate the workings of power in the modern organization of society and space.

Reorienting the Orient

Drawing upon Mitchell's arguments, let us now try to tackle the ways in which the above elucidated mechanisms have been implicated within the Orientalist construction of the Other: the archetypal product of the metaphysics of representation. In particular, I would like to look to the effects of this construction upon the modern traveler-tourist and his[4] attempts at (adventurous) forays 'beyond' the representations of the Other and into the spaces of the 'true' Orient, in the conviction (since still faithful to the logic of the world as exhibition) of being able to reach it at some point. Following the steps of such an imaginary journey and its appropriation of the spaces of the Other, we will try to shed light upon the mechanisms and the contradictions of the vision of the world-as-exhibition as a hidden technique for a theological construction of the world.

Our virtual traveler, armed with a series of representations of the Other gleaned from the variety of 'exhibitions' which have been paraded before his eyes in the course of a lifetime, departs in search of the reality which these above reenact. It is quite likely that our traveler's first contact with the ('real' or presumably 'real') Orient provokes bewilderment, confusion born of the disorder that it represents to his eyes. After some time, however, he begins to reorient himself, adopting some perspective able to convey a meaning to his explorations and to the myriad of stimuli before his eyes. If the exploration is that of an urban space, the perspective adopted could be that provided by a city map; that inspired by the description of yet another traveler (or travel-guide); or, better yet (since apparently more 'real'), that offered by a particular panoramic observation point.

This new 'orientation' allows our traveler to order the spectacle which assails his senses – and, in particular, his gaze – thanks to the reproduction (within this 'other' space, within the spaces of the Other) of the very logic of the world-as-exhibition: re-creating a dominating, external observatory, a perspectively 'meaningful' vantage point from which to view the object of interest. It is, of course, unsurprising that his comments and impressions atop the hillside or structure which acts as panoramic viewing platform faithfully reproduce the ensemble of preconceptions that have brought him to this 'foreign land' in the first place; a confirmation of the expected or, at most, a frisson of surprise (although surprise too is necessarily the fruit of a preexisting knowledge). In his intellectual and, inevitably, Orientalist ruminations, our traveler cannot but elaborate, critique, 'improve' a system of representations – the very same system which inspired his departure to the 'foreign land,' and to whose confines his explorations are, necessarily, confined.

This textual appropriation of the Other locates our traveler at the center of a strange optical effect: caught in the contradiction of reading a space which he considers 'real' (since he sees it, right before his eyes, in all of its pulsating vitality) as though it were a text, a map – in a continual search for the plan/meaning which it holds, since convinced, deep down, that there must, in the end, be some broader meaning, some order, some significance behind all that we see. In other words, in his immersion within the 'real' Orient, our traveler is constrained, despite himself, to reduce it to a text, to a representation of something else, to a reproduction of some symbolic but 'true' order. The traveler's gaze, moreover, presupposes a non-appearance, an invisibility within the reality which he is observing (thus, to the eyes of the observed, to the 'Orientals'): the ability to see without being seen. For it is precisely such invisibility (obviously impracticable) which confirms our traveler's collocation as the detached observer surveying the spectacle of the world. An attempt, more or less conscious, to reproduce the spatial conditions characteristic of the exhibition within which the status as subject-spectator grants our traveler an apparent freedom of movement between a diversity of observation points, without ever affecting/influencing the identity of the objects observed. This divide, this detachment (upon which, we will recall, the certainty of representation relies, according to Mitchell) is only reproducible with the concealment of the observer – and the naturalization of the center from which his perspective, his vantage point departs. In other words, such a detachment is only possible if the 'reality' before our traveler's eyes is imagined as a façade, as a spectacle – as the panorama of a whole which represents something else, which is endowed with some underlying meaning waiting to be revealed. It is a meaning, however, that is never present (and here lies the most intriguing aspect of this game), never accessible, but only represented (within the architecture of the city, within the 'typical' dwellings, within the faces of the natives, the goods displayed at the bazaars; in short, in that catalog of ambience and image within which our traveler has become accustomed to describing the world). The separation, the detachment offered by the view from the stage concedes, in fact, an innocence and objectivity to the modern observer and his analytical tools; an innocence which has been the focus of so much of social-theoretical critique.

Yet the paradox which underlies the logic of the metaphysics of representation remains concealed – and our explorer is not content with remaining invisible, or considering himself as such. Addicted to the logic of the world-as-exhibition, he still longs to immerse himself within that 'reality'; to 'enter,' at least for an instant, into that exterior whose descriptions he knows so well. Our traveler's need/desire both for a contextualized participation and a concurrent detachment are clearly paradoxical and theoretically unsustainable, though we can note that they are but the legitimate heirs of the equally paradoxical assumptions which have guided structuralist cultural anthropology. The anthropologist, in that optic, was to seek out knowledge of the Other: a search which was legitimized both by his immersion within the cultural 'reality' that constituted the object of analysis, by his participation in that culture and way of life – as well as, paradoxically, by his concurrent detachment as a researcher, a detachment which bestowed a 'scientific' value upon the knowledge produced.[5]

But let us return to our traveler. Trapped within this contradiction, he can only take refuge within the certainty of representation offered to him by the logic of the world-as-exhibition. For it is only within an exhibition, or within a world conceived as a pure exhibition of something beyond it, that these two positions – participation and 'objective' detachment – can be reconciled; a well-elaborated assumption in so many theme parks, shopping centers, and tourist itineraries, all of which accomplish such a reconciliation in an admirable fashion, playing upon the ambiguity between a passive and detached observation, on the one hand, and interaction/participation on the other. In fact, the certainty of representation, by reassuring us of the existence of some 'external' reality, also serves to shelter us from the anxiety that interacting merely with representations of the world could provoke.

How has modern logic succeeded in founding its 'regime of certainty' upon a similar paradox? How is it that we have forgotten this logic's founding act of 'faith,' its fundamentally *theological* constitutive principle which underlies the entirety of the theoretical apparatus that has allowed for the modern appropriation of the world? How is it that we have mistaken the 'special effect' of our relationship with reality for reality itself? What mechanism has allowed for the transformation of an ontological problem into an epistemological question?[6]

Let us leave our traveler to his own devices for a moment and let us attempt, rather, to investigate that regime of certainty which permits him to observe the spectacle of life from his window on the world. What if we were to invert the argument developed above? Following the Cartesian logic of the world-as-exhibition and thus the presumed distinction between our representations of the world and the world itself, we can note that any representation, any model holds as far as it provides the best possible approximation of an 'external' reality. Therefore, if within the spaces of the exhibition we find the representations of places, cultures, cities, it follows that such places, cultures, cities certainly *exist*. Better yet: the faithful reproduction of an 'exterior' only confirms the univocal and essentially stable relationship between the reproduction and thing reproduced. To the extent that the 'external' world is 'representable,' it is cognitively conceivable; it exists objectively as the referent or meaning for the signifier that stands before us. If we believe, following Saussure, that there exists a clear distinction between sound-image and meaning (thus, signifier and signified) – or, rather, that every signifier corresponds to a distinct meaning/signified, and that the power of this signifier rests within its ability to establish a stable and univocal relationship with its signified, what follows is that within the logic of the world-as-exhibition, every representation necessarily corresponds to a reality – this latter rendered 'real' precisely by the existence of its representation.

In this logic, therefore, only that which is representable exists, while that which is not representable, strictly speaking, does not. The space of the 'real' thus finds itself enclosed within the order of the representations which narrate it. According to the logic of the 'world-as-exhibition,' it is not the representation which relies upon the thing represented, it is not the map which depends on the territory but, rather, the opposite, even if this 'special effect' is well concealed by the veils of Cartesian certainty provided by the metaphysics of representation. The system of representations adopted to narrate the world through the reconstruction of a myriad of (more or less) material 'exhibitions' does, nevertheless, rely upon a certain logic, certain categories, certain languages and, above all, a very distinct set of perspectives. The certainty of representation thus confirms the existence of a reality which is revealed to us *only* through the conceptual cages that *our* system of representations has constructed in order to narrate this very reality! The 'special effect' we alluded to above, then, consists in the cancellation, in the forgetting of the original 'sin': the naturalization of the very specific nature of the meaning assigned to the representation which allows us to 'order' reality itself – while convinced, all along, of revealing, reflecting this very reality; of re-presenting it.

In consequence, if no other reality exists outside of that which is representable within our categories, all that which has not yet been included in our classifications or traced upon our maps *simply does not exist*, or *appears to our eyes as disorder, chaos, as an order waiting to be uncovered, to be 'put in order.'* The very idea of a disordered space (disordered since an untranslatable representation of some underlying order which, after all, must exist) only confirms, in the negative, the existence of order. In other words, according to this logic, the mapped space (if the map is 'properly' constructed) should reflect the order of the territory; the 'background noise' excluded from the map appears as such

only because it has not *yet* been mapped, though it is certainly mappable in theory. The problem of the certainty of representation is thus
transformed into an, essentially, technical question. The model can thus
be perfected in order to succeed, one day, in capturing even that background noise, and to collocate it within some order which will, finally,
grant it meaning. Here lies the triumph of cartographic reason: the colonization of the world enacted through the imposition of a spatial order
which is, inescapably, exhaustive, all-comprehensive. This spatial order,
measurable since detached from the 'things' which it contains, thus
becomes a neutral, abstract – and therefore innocent – space.

The metaphysics of representation unleashes its power precisely
through this appearance of innocence, of objectivity; through the
exhaustiveness of the spaces which it manages to create. The 'forgetting' of its founding act and the extraordinary ease with which it conceals this oversight through the colonization of the minds and bodies of
individuals (as Mitchell maintains) constitute the winning weaponry of
the modern vision of the world, of society, of space. Yet if it is here,
indeed, that the secret of the power of modern discourse lies, is its declaration, its 'unveiling' necessarily '*post*-modern'? If the metaphysics of
representation binds us within a 'closed' reality, framed by the order of
the representations themselves, can we term '*post*-modern' the rejection,
the refusal of the founding assumptions of this metaphysics (that is,
the refusal of a necessary distinction between representation and
thing/reality represented)? Not only: having rejected the logic of the
metaphysics of representation and having 'opened' its spaces, can we
formulate genuinely 'other' discourses, discourses able to transcend
this logic which are not reducible to analogies of the past 'order'? Can
geography, the description of the world *par excellence*, provide a truly
alternative vision? Will such be the task of postmodern geography?

Keeping in mind the theoretical elaboration of the mechanisms of the
metaphysics of representation, let us turn for possible answers to an
analysis of the classic geographical representation – that is, the map –
and the very particular relationship that this latter has instituted with
its traditional referent (the territory) within modernity.

The Map and the Plan

In his elaborations of the 'philosophy of the thing', Mitchell relates the
tour of Rabat organized by Marshal Lyautey, the colonial governor of
French-occupied Morocco during the early years of this century, for a
group of French engineers and journalists. I draw upon Mitchell's narrative of this tour as it highlights exceedingly well the ways in which

the mechanisms of the world-as-exhibition frame the reading of urban space (and particularly that of the European colonial city).

Citing the words of one of the participants, the writer André Maurois, Mitchell describes how Lyautey explains the idea behind the new colonial capital, "the philosophy of the thing," as he calls it, by pointing out how the buildings should be read "as a fan," with, at its center, the colonial Administration, and beyond the Government Ministries, "placed in the logical order." It is this preordained order that allows Lyautey to even collocate and describe the as yet unbuilt Finance Building, representable since destined to occupy its proper place: "This, here, is the gap for Finance. The building has not yet been built, but it will be intercalated in its logical place" (Mitchell 1998: 161). After having literally exhibited/exposed the meaning intrinsic to each of the buildings making up the colonial city as part of a greater plan, of a grand project, Lyautey's tour terminates, quite meaningfully, at the kiosk with maps, for

> The colonial city was to be unambiguously expressive. Its layout and its buildings were to represent, in the words of the architect who built Rabat, 'the genius for order, proportion and clear reasoning' of the French nation. As a system of political expression, each building in the city seemed to stand for something further. (1988: 161)

The perspective of Lyautey's reading of the urban space is decidedly cartographic, as it presupposes a certain equidistance and externality of the observation point adopted; an externality which allows one to grasp the meaning of the whole, of the underlying plan of which the layout, the external aspect, and the function of those buildings and those streets are a material representation. It seems only natural, then, that the tour should depart from an imagined cartography – and terminate at the map kiosk. This curious circularity inherent to the modern reading and construction of urban space proposed by the tour of Rabat is worthy of further reflection. Let us proceed, however, along Mitchell/Lyautey's voyage through the colonial city.

The urban landscape which stood before the eyes of the spectator/viewer was a distinct distribution of surfaces and spaces. Yet since the city before them was a sort of experiment/exhibition of the linearity and functionality of modern canons for the design of the urban context, this distribution of spaces and surfaces appeared extraordinarily regular and orderly. The eye of the observer would thus perceive this regularity as a sort of material manifestation of a grand project, an overarching plan. The regular and geometric spaces of the colonial city would thus reveal a strong intentionality, a sense of order underlying the visible architectural expressions, while these latter were to materialize and at the same time represent the functional structure of that very same order. The spectator, 'formed' as he was by the European exhibitions, was thus wont to think that the impression of such a regular, orderly distribution was formed by two distinct entities: one spatial and material, the other nonspatial and conceptual – the project which stood behind the realization of that distribution. Thus, the buildings, the streets, the urban landscape writ large, on the one hand; on the other, the plan, the deep meaning necessarily represented by that landscape, by that architectural ensemble. This interpretation, stressed in Mitchell's narrative of the representation offered by the 'creator' of this plan, Lyautey himself, encouraged the European spectator (just as the present-day tourist) to view the spectacle presented by that urban space within a dual perspective: a perspective which kept 'the things' rigorously separated from their philosophy, from the thought which had conceived them – the city, in all its materiality, from the plan, the project, the map which had rendered it possible.

This rigorous separation between the city and the plan, between the materiality which stands before our eyes and its deeper meaning, has guided modern reason through its colonization of the world to this very day. And it is this very separation which has bred a cartographic reading of space, basing this latter upon a regime of certainty which, paradoxically, emerges from an astounding contradiction. In fact, cartographic reason's ability to mask and/or forget this contradiction can be traced precisely to the 'reality-effect' produced by the mechanisms of the world-as-exhibition. In the pages to follow, I will attempt to isolate this effect – pronouncing if not the death of cartographic reason then, at least, the theoretical unsustainability of that regime of certainty upon which it is founded.

As Mitchell himself notes, to the eyes of the European visitor to Rabat, such a division of the world in two appeared absolutely meaningful and quite natural. Accustomed to reasoning in Cartesian terms, those of a Mind which ordered external reality through a series of images, accustomed to thinking/seeing the world through the exposition of representations offered up on modernity's plate, it is no surprise that Lyautey's guests were quite content to accept the dualistic reading of the world and, therefore, of the urban spaces before them. The French engineers and journalists described by Mitchell were scarred by the very same logic that has formed our own traveler; though far from home, far from the spaces of the exhibition, they could only continue to interpret the world as though it were an exhibition. The very same 'rule' which decreed a clear separation between representation and reality governed, moreover, the construction of museums and zoological gardens, the organization of Orientalists' congresses, the publication of statistical reports and legal codes and, perhaps most visibly, the construction of department stores and all the surrounding material architecture of the city.

To the eyes of anyone traveling through the modern world (or better yet, through the spaces created by the theory of modern representation), things thus necessarily seemed constructed, arranged, conceived, ordered, and even consumed as signs of something else. A street, a panorama, a book, a piece of merchandise, appeared as mere objects or as an ensemble of objects organized, 'elected' to represent (in a sort of infinite exhibition) some meaning, some culture, some society; as material testimony of the existence of some idea, some more 'original' experience:

> The arrangement of buildings seemed to express the institutions and authority of a political power, Alpine scenery became the experience of Nature, articles in museums conveyed the presence of history and culture, words in Oriental languages represented an exotic past, animals in zoos an exotic present. Life was more and more to be lived as though the world itself were an exhibition, an exhibition of the exotic, of experience, of the original, of the real. (Mitchell 1988: 172)

The principle of the exhibition thus became the principle of the experience and the understanding of the world as a whole. It is for this reason that the careful ordering of the built architecture, of panoramic views, of the displays which offered up goods for sale, of the individual travel experiences through this world, attempted, at all costs, to reduce everything to a mere representation of "something more real beyond itself, something original outside" (1988: 173); a conceptual, abstract realm where 'reality' would reside.

The so-called 'reality-effect' of modern Western thought relies precisely upon this absolute and exhaustive distinction between 'mere things-in-themselves' and the 'real' meaning for which they stand, thus reducing the interpretation of the world to a univocal relationship between signifier and signified. Keeping the above reflection in mind, let us now return to the two issues that I highlighted in my introductory remarks: the relationship between the description/narration of space and its construction on the one hand, and that between society and space on the other.

The Map and the Truth

I would like to begin with the notion of the 'world-as-exhibition' itself – and the clear-cut distinction between representations of the world and some 'external' reality upon which these purportedly rest. It is by virtue of this distinction that the modern reading of the world can be understood as an infinite series of exhibitions; as a display whose materiality recalls another, symbolic-conceptual dimension. If we accept this conceptualization, then territory itself can be interpreted not so much as 'reality' but rather as the material manifestation of 'something else', of the idea of Nature, of culture, of society, of the city etc. The territory thus becomes a sort of 'façade' which represents some 'external' meaning, some plan, some project, some series of 'orders'; all these embodied in its visible elements and its organization.

What about the map, then? If, in fact, the territory is interpreted as the material dimension of some underlying truth, as the manifestation of an order, of a plan/project, we can then identify a whole series of consequences for the cartography which seeks to describe it. First of all, our map is no longer a representation of reality, but rather the representation of the representation of something else, of some 'real' original which belongs to the reign of meaning. Is the map, then, less credible, less dignified in its role as a useful and legitimate reading of the world? Curiously enough, this is not the case. Let us call this fictitious map 'Map A.' 'Map A' is a reproduction of a territory – a representation of a dimension that proposes itself as the representation of some project/plan, some meaning, some underlying order which cannot be made present/materialized but can only be (re)presented through the materiality of the territory itself.[7] If, then, there is some external meaning to things, some 'real' realm which transcends the materiality of the territory, such meaning, in the logic of the metaphysics of representation, comprises both the deep meaning *inherent* to things in themselves (the *genre de vie*, culture, tradition etc.) as well as the (intentional) *assigning* of meaning, thus the outcome of a process of signification (the plan of Rabat, the order of the State, the territorial plan etc.). These two meanings (which appear to be separate, since the former is transcendent while the latter is intentional) happen to coincide within the metaphysics of representation as they occupy the very same space: that immense 'exterior' which corresponds, at base, to the domain of the 'real,' the domain of significance.

Imagining society, culture, history as 'external realities' with the capacity to confer meaning on the representations of the world (and thus also to the territory itself) means collocating them at the same level of the colonial plan for Rabat: readable exactly through its process of signification of the urban space. The order which 'appears' within Rabat's urban structure is clearly cartographic in nature (something which appears quite obvious to Lyautey and his guests); a cartographic order which, by imposing a 'true' meaning, allows for the establishment of a univocal relationship with the territory which should serve as its representation. In fact, turning this argument around, if every representation necessarily embodies some further meaning, then every territory (which corresponds to a representation) relies upon some sort of hidden cartography, or a series of such cartographies which have contributed to its signification, to its ordering. Rabat's plan is the *exterior* which conveys meaning to the urban spaces of the new colonial capital. But, as a plan, a project, it is clearly an intentional act, a genuine product of the cartographic colonial reason; the materialization, elsewhere, of the European urban order. It is precisely in this formidable ambiguity between the essentiality and the transcendental character of the underlying order on the one hand, and its concurrent intentional nature on the other, that the power of the metaphysics of representation lies hidden. No distinction, then, appears between plan and culture, between project and nature, once we find both amalgamated in the 'external' reign of meaning. And it is upon this 'magic of significance,' upon that certainty attributable to it, that the modern construction of the world rests.

Let us term this hidden (and thus to be unveiled) cartography of meaning 'Map B.' In the logic of the world-as-exhibition, we will recall, Map B is visible only through its material manifestations (thus, the territory itself), although *it* is the real referent, the only reality accessible, yet one we can never 'reach,' resigning ourselves to its mere representations.

What does this certainty consist of? Or, rather, what are the workings of that mechanism which allow us to consider that plan/project (Map B) as a 'reality,' as an original? Within the Cartesian logic of the world-as-exhibition, our Mind is, curiously, both an 'external' (to the world) and 'internal' (to us) observer; an observer seeking to represent, in an ordered/orderly fashion, the reality 'outside.' Yet if the order of these representations must, necessarily, reflect the order of that 'external' reality, if it is to be useful/meaningful (thus the search for exhaustive, 'complete' models which could provide a faithful mirror of reality), then we must first accept the very *existence* of such a representable reality (in our case, the existence of Map B); not only, we must also accept the idea that for this symbolic/projectual dimension to be 'real' it must, perforce, be representable and thus be *reflected in the materiality of the world*. And it is for that very same reason that we will consider 'real' only that which enters into the order of our representations. In cartographic terms, then, Map B (the plan/project, the underlying order/meaning of things) can be considered 'real' as long as it 'fits' into the order of Map A.

What appears here is a curious coincidence; a coincidence which allows us, finally, to provide justification for the affirmation that within modern cartographic logic, the description/narration of a territory contributes to the latter's construction – and vice versa. We can also hazard, at this point, the rather bold assertion that within the closed spaces of the metaphysics of representation, *the map coincides with the territory*; representation and thing represented end up being one and the same. In fact, the territory, in the cartographic spaces of the world as exhibition, exists only as a reflection of the plan/project, of the underlying conceptual dimension that is Map B. To exist, however, this territory must be capable of being ordered and thus, necessarily, representable within . . . the order of the map (Map A). The description of the territory (Map A) and the conceptual 'reality' or meaning to which the latter alludes/refers (Map B), therefore, tend to coincide. If the above is true, however, then the territory (within the logic of the world-as-exhibition) acts as a strange sort of interface between the two cartographies, between two representations of the very same nature which claim one to be the mere description of reality, while the other, reality itself. Yet reality, the only reality possible within this closed space, is that of the map; that of our representations which are not more or less faithful reflections of reality but are reality itself. The territory thus becomes a double-sided mirror, reflecting the same macabre image.

Here lies the 'secret' of the colonization of the world so aptly described by Mitchell. It is here that we come face to face with the iron-clad logic of cartographic reason, a logic which recognizes the existence of the territory in the only form in which it is capable of conceiving it: as the representation of a plan, of a project; essentially, as the representation of the cartography which has produced it. Should this form not be evident at first glance, with the territory appearing indecipherable and indescribable, it must, immediately, be granted meaning which recognizes the existence (obscure at the moment though certainly present) of a (dis)order; a (dis)order created by the very same representations which seek to unveil it. Here rests the certainty of the world-as-exhibition. Here lies the true power of the metaphysics of representation – a power which, by subsuming *all of reality* within the order of its own representations, colonizes our geographical imagination(s) and thus our vision of the world. It is a particularly effective power for it is "never presented, . . . only ever represented," to cite Mitchell (1988: 179), and thus permits us to assail only its ephemeral representatives, its representations – but never to attain and to unmask its deepest logic.

Can we term 'postmodern,' then, the declaration and isolation of this mechanism of power? Can we formulate a geographical praxis capable of eluding its omnicomprehensive logic?

(Post)Modern Space

At this point we can tackle the third problem highlighted in the introduction: that of the relationship between postmodern philosophical reflection in geography and the parallel construction/codification of postmodern spaces. We should begin with some general considerations: if the construction of modern spaces (and their description/narration, as we have seen) follows the logic of the world-as-exhibition, then that of postmodern spaces, we can hazard, should reject such logic. What 'code' would allow us to resignify spaces in postmodern terms, however? Can we imagine such spaces?

First of all, such spaces should be ones within which the fatal divide between representation and reality does not hold; thus spaces which are not worth merely what they represent. The construction of such spaces should, therefore, presuppose the refusal of a cartographic codification of the relationship between a given space and its respective meaning. Spaces, then, which are unmappable; spaces which do not necessarily reflect a 'something else,' something symbolic, 'deep,' and true; spaces which, above all, remain 'open' to the multiplicity of interpretations that their actors elaborate in an endless dialectic. Spaces definable only within a continual recontextualization, as any attempt to 'freeze' these within a cartographic frame, within a 'certain' description, would transform them immediately into modern spaces. But if these, indeed, are the conditions which could designate a space as postmodern, can such spaces truly exist? Or can we only speak of a so-called postmodern awareness of space, an analytical approach which does not attempt to 'tag' spaces by mapping them? In this perspective what would a postmodern city be? A space shaped by the above described canons, or merely a city which 'allows' for a multiplicity of interpretations of its very spaces?

Let us try to provide answers to some of these questions by turning to what is considered by many an archetypal postmodern space: the Disney theme park. The choice is hardly accidental: the Disney amusement parks embody, after all, an extreme version of those exhibition-spaces from which Mitchell's analysis of the world-as-exhibition departs. The Disney-spaces, with their extraordinary emphasis on the reproduction of historical and geographical alterities, are a present-day incarnation of the exhibition spaces of the late 1800s which respects the latter's canons to a T: the extreme attention to detail, the construction of the exhibition space around a common 'center,' the collocation of the visitor/spectator as the occupant of this central point. Not only; for quite some time, theorists have pointed to a 'Disneyfication' of social space writ large (perhaps most evident in the North American context), that is, the increasing adoption of the Disney rhetoric in the construction of socially-strategic urban spaces, with its strong emphasis for the semantic reconstruction of determinate atmospheres as well as a certain thematic recontextualization. This process can be observed both in projects for the revitalization of urban waterfronts, as well as within the latest generation of shopping malls which attempt to recreate within their walls the traditional places of European urban socialization. Within these reconstructions, particular attention is paid to the role of the façade, to its evocative power, to its thematic coherence, this latter often lacking any relationship to the functionality of that same space. I will focus, in fact, upon this lack of a direct and univocal relationship between the façade of the so-called postmodern space and its functionality (its *raison d'être*, as it may be), for it is a point which may help furnish some answers to the questions posed previously. Following the temptations of the postmodern spirit, I would like to propose two possible readings of this fragment of the contemporary urban environment – although many others are certainly possible.

In a first hypothesis, the Disney-spaces could be interpreted as simply a radical(ized) version of the world-as-exhibition itself, with its continual (even if not altogether serious) reflection of the outside world, of some conceptual realm which it does its best to reproduce, following faithfully in the tradition of the exhibitions of the 1800s. Successful gated communities, theme parks, waterfront façades, and revitalized 'historical districts,' in fact, attempt to materially re-create a series of representations of the distant past or present with painstaking attention to the 'atmosphere' of the theme reproduced. The distant and the Other are thus narrated through the techniques of the exhibition, their reproductions imagined as the representations of the 'culture' of a distinct society; spectacularizing, in other words, the distinction between a series of representative objects and a corresponding series of containers (culture, society, history etc.) which exist 'out there' and can thus be narrated within the order of the exhibition.[8] The Disney-spaces, in this optic, appear to coincide with those of the world-as-exhibition; appear simply as an extreme version of this latter. Accepting the *logic of the descriptions* of the Other proposed by Disney's reproductions (though not necessarily the descriptions *in themselves*, we should make clear), *any* alterity can be collocated within a specific 'culture,' a certain 'society,' and thus be rendered describable, classifiable. It is such a classification, after all, which allows us to capture its 'typical traits,' its 'character,' the material manifestations which encapsulate its specificity; it is with this conceptual operation that we virtually enclose the Other and the world within the very same categories ordained to represent it within the theme park universe. The fact that Disney's frivolousness and liberty of description can be accused of artificiality only favors the game: it is precisely in the ambiguity between 'reality' and fantasy that the Disney spaces confirm that vision of the world which rests upon a catalog of all its manifestations within some container, within some 'external' order. The world thus becomes recognizable, understandable through an analysis of the very categories which (seek to) contain it. Yet if we are to accept this interpretation, then we certainly cannot speak of postmodern spaces but, rather, of an extreme radicalization of the modern discourse, perhaps even a hypermodernity. All that would remain for the postmodern would be a possible reinterpretation of these spaces.

But there is a second reading which can be proposed. In this vision, the Disney-spaces could be considered truly *post*-modern spaces, since they embody the collapse of the signifying chain. In other words, the very fact that the landscape of these spaces is a self-admittedly reconstructed façade which makes explicit the absence of a necessary relation between the materiality and the functionality (the plan) of that environment, could give credence to the idea of the existence of spaces which cannot be described according to the logic of the metaphysics of representation; this latter would be, in such a hypothesis, rendered useless, precisely due to the free and unrestrained use of images that characterizes the world of Disney. This is certainly an hypothesis worthy of attention, for the Disney-spaces, by taking the modern game of signification to its extreme, actually play upon an extraordinary paradox. On the one hand, through their use of simulacra and simulated landscapes, these spaces evoke the idea of elsewhereness, of some world 'beyond,' of some deeper meaning which can be attributed to the objects reproduced 'within.' On the other, all such reproductions appear as (somewhat) veiled pretexts, recalling 'external' reality in an ambiguous and ironic fashion; tending quite often, in fact, to strip this external reality of meaning, rendering it much less 'interesting' in comparison with the scintillating spectacularity of its 'internal' representations. It is this marvellous ambiguity that allows the Disney-spaces to deconstruct cartographic readings of the world, mocking them with its liberal construction of images of reality – and its deliberate 'forgetting' of the referent. These spaces, in fact, have often been assailed by modern(ist) critics as 'artificial,' as mere representations of representations (of the materiality of the world), as grotesque parodies of the 'map' carried out for commercial reasons alone. Viewed in this optic, however, Disney-spaces could symbolize a refusal of the modern logic of the world-as-exhibition.

Opening up the Disney-spaces to a double reading is intentionally provocatory, hoping to legitimize the interpretative pluralism of a space whose very success and role as a cultural product are based *precisely* within the ambiguity of its signs. In such a pluralist vision, we can propose that a space can be modern or postmodern according to the interpretation and the use made of it by its protagonists/visitors/spectators. In other words, assigning a 'strong' meaning to a space, determining in a univocal manner its very identity and essence, requires a decidedly cartographic – and thus modern – approach. It is only by envisioning spaces as flexible, in continual transformation, in a continual recontextualization of their meanings, that we can approach a postmodern sensibility, a postmodern awareness of space.

It makes no sense, then, to speak of spaces that are postmodern in essence, as such an affirmation of identity would, inevitably, fall into the trap of the modern logic of the construction and reading of spaces. The Disney-spaces, then, are decidedly 'modern,' though with a variety of tonalities, both for those who consider them 'false' and 'mystificatory' (since such a reading is predicated upon the assumption that other, more 'true' and 'honest' representations of the realities recounted within the park exist), as well as for those who place their trust within the Disney representations as not-too-serious though plausible narratives of the world; a world describable only within the exhaustive categories proposed by these very representations (culture, society, national character etc.). They are postmodern spaces, on the other hand, for those who discover within them the material reconstruction of the collapse of the signifying chain, and thus a sort of liberation from its cage of truths, explanations, and exhaustivity. Or, for those who perceive them as but one of the many possible descriptions of the world, an experience comparable to a tour of an unknown city or some virtual voyage. For this latter 'interpreter,' the relationship with the exhibition is but a game, in the conviction that the world itself, in all its material expressions, is but a never-ending game of citations.

It would be important to note, nevertheless, that even the very attempt to identify a clear distinction between a 'modern' and a 'postmodern' interpretation or sensibility of space risks falling back into the constraints of the dualisms of modern logic, operating, as it does, within the terms of an acceptance or rejection. How can we escape these omnipresent terms of comparison, however? Is this but proof again of the pervasive logic of the metaphysics of representation, a logic which resurfaces, time and time again, in all of our considerations of the world?

Beyond the Map

It is again the metaphor of the modern traveler that can, perhaps, allow us to move beyond the binding ties of cartographic reason. We had left our traveler gazing out at the spectacle of the world from the high ground of his panoramic position, Orientalist map of the Orient that stood before him in hand. We can, for argument's sake, imagine our traveler in Cairo (a destination which was to inspire so many of the Orientalist dreams of the modern European spirit), searching, in a more or less conscious fashion, for some Ariadne's thread capable of giving sense to the complexity of the Oriental spaces before his eyes.

Several options lie before our traveler. He can certainly tackle the city armed with the readily-available tourist cartography; that is, by adopting a strictly 'closed' reading of the Egyptian capital, assuming its perspectives, its (more or less exotic) monumental vision, its historicist interpretation, its search for – and reconstruction of – 'Oriental' atmospheres. Traveling across this Orientalist space, which assumes the shape of a veritable territorial *strategy*, our traveler can choose to accept it, in uncritical fashion, as simply a plausible – and consumable – interpretation of Cairo as a representative fragment of the Orient. He can, however, also choose to adopt a series of what de Certeau (1984) would term *tactics*: utilizing the tools and representational codes of the tourist cartography to undermine this latter's credibility 'from the inside,' unmasking its perspectives (by, for example, photographing the Sphinx from 'the wrong side,' from behind, and thus exposing Cairo's rampant building speculation, rapidly encroaching upon the park grounds); exposing, in other words, the mechanisms which allow the strategy to capture Cairo's spaces within its grid of power, within its order. An interesting and creative exercise, certainly; one whose object would lie *not* within a deconstruction of the strategy's 'scheme of things' (in order to attempt to impose another, 'better,' order) but which would, rather, attempt to unmask the partiality – and the underlying dynamics of power – of this very strategy *within its own spaces*. Yet both above perspectives are still inescapably bound to the tourist cartography, caged within the closed space that this latter attempts to impose upon any interpretation of Cairo's urban context.

A seemingly revolutionary and (admittedly) 'alternative' attitude is that of the traveler-spectator who 'detaches' himself from the map, wholeheartedly rejecting the tourist reading of Cairo: plunging into the disorder of the urban environment, abandoning well-trodden tourist routes to lose himself in the labyrinthine passageways of the medieval Islamic city, in the smells and sounds of the 'local' marketplaces, in the street-side coffee-shops frequented by Egyptians alone. Determined to flee the constraints of the tourist cartography, our 'alternative' traveler seeks to reach out and touch the true Orient, to experience the genuine atmospheres of the 'real life' of this pulsating city, to capture within his mind a 'truer' image of the spaces before him. Yet he too remains bound to the map, for his strategy is but a rejection of the imposed model in a desperate search for a 'better' map than the one provided by the tourist imaginary.

What attitude, what perspective, then, would allow our traveler to 'open' the closed spaces of modern cartographic logic in which he finds himself ensnared? What reading of Cairo would allow him to escape from the dichotomy inherent in any search for a distinction between 'Modern' and 'Oriental' spaces? A reading, perhaps, which would take our traveler into and across Cairo, still with map in hand – a map capable of giving meaning and coherence to his explorations – yet now also armed with the awareness that the cartography guiding his voyage is but one among thousands, but one of the infinite number of paths possible within that space, but one of the infinity of possible descriptions of that very territory. Rendering explicit its partial and positioned nature, his reading will thus be able to face 'reality' – that is, that unforeseeable, unpredictable outcome in continual transformation that the dialectic between a multiplicity of cartographies (his included) produces within that space. His interpretative path will thus be an explicit one, declaring both its intentionality and its discourse; an essentially open path, since the outcome of its relationship with reality shall remain entirely unpredictable. Unpredictable, for a multiplicity of other paths are concurrently weaving their way across that very background, crossing his own and giving life to a reality which changes precisely in function of the infinite number of recontextualizations that each description of space enacts. In this fashion, our traveler's voyage can become an itinerary traversing an infinite series of contexts in continual evolution and redefinition; contexts which he can only attempt to 'capture' through the theoretical filter that he decides to adopt, a choice recontextualized, reinterrogated time after time. Such is the only possible relationship with reality conceded to our 'post-modern'[9] traveler. A relationship that is partial, contextual (or, better yet, continually recontextualizeable), but one that is also extraordinarily open and unpredictable.

By disowning the logic of the world-as-exhibition, our traveler's vision of the world around him will be an 'other' one: he will no longer look to the map to reveal some deep and hidden reality; no more envision space/territory as the material manifestation of some greater truth, some grand plan. It is his description, along with the multiplicity of other descriptions of that very same space, in an unending dialectic of power, that will contribute to shaping that space, that territory, in practice; within a discursive-operative logic which forges it continually. The philosophical reflection on the significance of geographical discourse, on the description of the world and of its spaces and on the construction of these latter thus come together; theory and praxis becoming one and the same, acting (together) to create contexts; description becoming the conceptual/operative tool which allows us to cognitively 'capture' the multiplicity of spaces/territories that reality presents to our explorations.

It is the apparently paradoxical play between the theoretical *closure* enforced by the logic of our privileged description – and the simultaneous *openness* of this latter to continual recontextualization – that precludes the assigning of definite meanings to determinate spaces. Such a dialectic between *continuity* (our chosen path, our privileged description) and *discontinuity* (its inevitable recontextualization) closely resembles the Derridean game of words and signs. If we do, in fact, repudiate the existence of a separation between sign and meaning, between a specific space and the plan which sustains it, we can then, perhaps, imagine space as an incessant play of continuity (sameness) and the renegotiation of meaning (difference): the continuity deriving from a sort of *memory* of the endless 'negotiations of meaning' between the competing descriptions which have contributed to the construction of that space and its 'history' – a memory which, just as the *trace of sameness* of the Derridean sign, allows for the (always temporary) granting of meaning and its communication. The renegotiation of such meaning is incessant, however, as is the production of alternative pathways across that very same space, pathways which, certainly, cannot abstract themselves from the memory of past negotiations of meaning, but which never naturalize such memory as essence, as character; never attempt to cage it within the bars of cartographic reason. Meaning thus emerges from this paradoxical quality of *sameness* and *difference*: just as a word '*always happens to be just the same,*' as otherwise it would fail to communicate any meaning, '*only different,*' since incessantly recontextualized, so too the 'meaning' of a determinate space *always happens to be just the same,* framed by the traces of past negotiations, *but only different,* transformed in practice by those who live and describe/narrate it.

Let us return, one final time, to our traveler's dilemma. The 'other' path which we have traced allows him not so much to escape from the modern logic of the world-as-exhibition as to isolate it, to open its spaces, to render banal its pretensions. In such a vision:

everything occurs as the trace of what precedes and follows it, nothing is determined as the original. Nothing stands apart from what resembles or differs, as the simple, self-identical original, the way a real world is thought to stand outside the exhibitions. There is no hierarchical order of the imitator and the imitated, as in an exhibition or any other system of representation . . . There is no simple division into an order of copies and an order of originals, of pictures and what they represent, of exhibits and reality, of the text and the real world, of signifiers and signifieds – the simple, hierarchical division that, for the modern world is 'what constitutes order.' (Mitchell 1988: 61)

The spaces that our traveler will encounter in his explorations through Cairo's streets are now free to assume a new, open, meaning. These will no longer appear disordered, since an order, a true order from which to distinguish a disorder, will have disappeared from his sights; the city will present itself to him in all its complexity – the fruit of ongoing processes of signification and the trace of past negotiations of meaning. Spaces, then, which are never mappable, never definitively representable, only liveable – only 'presentable' within his (privileged) geographical praxis. Our traveler could even write a narrative account of his voyages: yet any such story shall be but an account of the relationship between his cartography, his description – and the multiplicity of recontextualizations to which he had contributed, in his role as narrator/constructor of the world, within the course of his adventures.

Strategies . . .

My initial 'temptation(s)' offer, perhaps, a strategy of sorts. The two broad questions that have guided the reflections within this chapter focused upon, respectively, the by-now common assertion according to which geographical representations not only narrate reality but also contribute to its construction, and the idea of space as the product – and producer – of the social dynamic. It is with these in mind that I attempted to overcome the apparent disjuncture between the theoretical-philosophical reflection on the role of the postmodern in geography – and the analysis of the presumed postmodern spaces. In my forays into the above themes, I turned to the metaphor of the voyage – in the awareness, perhaps, that it was only in a voyage beyond the 'cages' imposed by modern logic that my second 'postmodern temptation' lay. A voyage in which we would attempt, armed with the conceptual categories of the world-as-exhibition, to somehow isolate, codify, the closed spaces that cartographic reason has attempted to impose on the world and our ways of conceptualizing it.

My explorations have been guided by the critique of the metaphysics of representation developed by Timothy Mitchell in the pages of *Colonizing Egypt* and his trenchant dissection of modern reason. Following Mitchell's theoretical trail, I have attempted to describe the 'hierarchy of truth' which lies at the bases of the modern colonization of the world, analyzing, in particular, the relationship that the map had established with its presumed referent: the territory. Deconstructing the assumptions that have held up the logic of this 'regime of certainty,' and attempting to reveal its so-called theological effect – that is, the ontological, not epistemological nature of the necessary separation between representation and reality – I suggested that within the closed spaces traced by cartographic reason, the map coincided, for all extents and purposes, with the territory itself. Or, perhaps more accurately, that reality – that conceptual realm which the territory described by the map should recall – was nothing more than yet another map.

In order to propose some possible 'ways out' of such anguishing circularity, I highlighted a number of imaginations of other, '*post*-modern' spaces; spaces constituted within logics other than those bound within the modern, cartographic 'cage.' The first such were the Disney-spaces considered the archetypal 'postmodern spaces' by many geographers. An initial verdict in this case seemed to suggest that the very same spatial environments, the very same spaces, could be considered either modern or postmodern, according to the particular analytical approach privileged – just as such spaces could be lived and experienced in a modern or postmodern fashion according to the predilections, to the 'sensibility,' of their protagonists. Yet the very dichotomy – 'modern'/'postmodern' – and thus the consequent need to collocate our vision of space within but *one* of these categories are still, inescapably, bound within the cognitive closure, the cognitive 'cage' of the (modern) metaphysics of representation.

My conclusions are more of a departure than an arrival. What I would like to argue is that it is only by abandoning all pretenses of crafting a 'fixed' description of space/territory (as though this latter were some distinct entity endowed with a certain meaning that we can, somehow, 'reach,' discover, and thus codify) that we can attempt to 'open' the closed spaces inscribed by the cartographic reading of the world. Mimicking the Derridean game of sameness and difference, we should seek readings of space/territory that are not only partial and positioned, but also always 'open,' since infinitely recontextualizeable. Within such (*post*-modern?) readings, spaces and territories assume entirely novel meanings – no longer simply the material manifestations of some underlying essence, of some identity to be revealed, they become the scenario and the product of an infinite recontextualization; a context in continual evolution and transformation, guided by the multitude of descriptions that contribute to its narration and construction.

Geographical praxis, in such a reinterpretation, thus frees itself from its time-honored search for truth, from its search for some (external) meaning that spaces/territories should reflect; it becomes, rather, an instrument through which we can both narrate the world as well as construct it – in its continual process of negotiation and articulation with a multiplicity of other praxes. Reality can thus only be 'encountered' in its lived experience; in the fleeting moments within which we contribute to its narration – and thus to its transformation, to its continual recontextualization within the cartography that we have chosen to adopt at that particular moment in time. A *post*-modern geographical praxis, a praxis which isolates and opens the modern method (of representation), could thus lead us into a new exploration of the world, leaving inevitable traces of our passage in an infinite game of negotiation of meaning with multiple other discourses, multiple other geographies. Without mistaking, this time around, Ariadne's thread for the world itself.

NOTES

1   In the contemporary Disney-esque variant this realism has been brought to new heights with the extraordinary 'reproductive' capacity of the parks' imagineerings.
2   As Mitchell (1988: 12) notes, in the modern European city everything was "organized to represent, to recall like the exhibition some larger meaning. Outside the world exhibition, it follows paradoxically, one encountered not the real world but only further models and representations of the real."
3   Here, Mitchell (1988: 18) cites Walter Benjamin's (1978) observation that exhibitions "open up a phantasmagoria that people enter to be amused ... They submit to being manipulated while enjoying their alienation from themselves and others."
4   The choice of a male subject here – and throughout the piece – is intentional: the modern 'traveler' and his construction of the Other can only be understood in the perspective of a dominant and dominating male gaze, as feminist and postcolonial critiques have rightly stressed.
5   Here, we should be infinitely grateful to James Clifford (1986) and the Santa Fe group for having exposed this game with extraordinary lucidity.
6   Franco Farinelli's (1990) work offers a fascinating reconstruction of this 'forgetfulness' within modern geographical thought.
7   The geographical tradition has accustomed us to describing territory as something which embodies a series of meanings, as the reflection and product of a given society, as the mirror of the relationship between two dualities: nature and culture, the physical environment and the (human)

community which inhabits it. It is upon a similar assumption that the Vidalian concept of the *genre de vie* is founded, as is, in more general terms, the very idea of region, if we conceive of the latter as an organism endowed with some sort of personality, some series of objectives, some ends (such as, for example, the maintenance and possible reinforcement of its own autonomy and identity).

8 This genuinely modern tendency of 'displaying' the world is perhaps best codified within Disneyworld's EPCOT 'World Showcase.'

9 '*Post*-modern' since 'other than modern,' attempting to open the logic and the spaces of modernity; the hyphen is meant to stress this 'opening' rather than a comparative/dichotomous relation, *viz.* 'modern'/'postmodern.'

REFERENCES

Clifford, J. and G: Marcus, eds. (1986). *Writing Culture: The Poetics and Politics of Ethnography*. Berkeley: University of California Press.

de Certeau, M. (1984). *The Practice of Everyday Life*. Berkeley: University of California Press.

Farinelli, F. (1990). Epistemologia e geografia. In *Aspetti e Problemi della Geografia*, ed. G. Corna Pellegrini. Milano: Marzorati.

Mitchell, T. (1988). *Colonizing Egypt*. Berkeley: University of California Press.

10

Paradoxes of Modern and Postmodern Geography: Heterotopia of Landscape and Cartographic Logic

Vincenzo Guarrasi

Environment and Culture

Carlo Socco's (1998) *Polysemy of Landscape* provides a definition of landscape that I have always found particularly useful: "all landscape, whether inhabited or not, is culture: that is, a means of signification and communication." It is a definition which rests upon a key assumption, for within it, the environment (or the 'physical world') is seen not as *external* to culture but, rather, incorporated *within it* through semiosis, the 'cultural act' *par excellence*. Our definitions of a 'cultural landscape' thus necessarily comprise the natural environment as well. Upon a world of 'natural' phenomena, phenomena which we strive to discover and to describe, we inscribe an artificial 'order,' that of 'culture' as a signifying system within which every material object is associated with a mental representation; every signifier with a corresponding signified. All manifestations of human spatiality are thus both material and mental objects – they are *signs*. And if we hold the above to be true, then space and culture become indistinguishable.

A Spatial Metalanguage?

The problem, if anything, lies elsewhere: that is, how do we inscribe another order unto a preexisting one? How can human societies act to modify existing cultural orders – and their accordant spatialities – in order to generate entirely new ones? In particular, how can a city or metropolis (in its etymological meaning of 'mother-city') create ever new orders within spaces that are already signified, already 'artificial'? I would claim that this capacity for crafting ever new spatialities which allows us to introduce endless new orders to the world around us is, inescapably, tied to our 'metalinguistic' abilities: abilities which allow us to move ceaselessly between the signs of object-language and the signs of metalanguage.[1] Our ability, then, to *resignify*: our ability to transform a sign into the signifier of something else. It is here that the innovative and dynamic-creative character of human semiosis lies – and landscape is but one of the multiplicity of metalinguistic acts which allow us to resignify the world. More than an 'act,' however, landscape is perhaps best envisioned as discourse or even a form of rhetoric.

These terms should not be used loosely, however. Coming from a background in anthropology and having studied the semiotics of culture, I have always been wary of the uncritical adaptation of a variety of general concepts to 'geographical' questions. In particular, I have always questioned the utility of the textual metaphor within geography. We may have become masters of literary analysis, that is true, but the temptation to transpose already proven heuristic devices to fields other than literary criticism might prove a dangerous one, forcing the examination of human societies and spatialities into an analytical grid which does not befit it. Or not completely, anyway.

What do we lose along the way when we choose to analyse territories, landscapes, or environments as 'texts' (Duncan 1990)? What do we lose of the corresponding 'polysemies'? Each of us expresses himself or herself through language, but we also live and act *in space* – and it is *both* through linguistic *and* spatial practices that we grant meaning to our lives and to the world. If in our analyses we simply conflate the linguistic dimension with the spatial one, we may easily jeopardize our understanding of the delicate equilibrium between these two dimensions in shaping/guiding human action; a simplification of even graver proportions should it occur not on the level of object-language *but of metalanguage itself.*

The Notion of Heterotopia

We can define territory as a complex of material objects located in space. Or, as it is often defined, as a complex of places, that is, of spatiotemporal phenomena. Can this spatial dimension be fully analyzed through a 'semiotics of narration'? Or is it partially obscured, given this latter's privileging of the temporal dimension over the spatial one, of the diachronous over the synchronous? Certainly, if we hope to escape from

the temptations presented by the narrative text, we need to arm ourselves with the appropriate tools. And it is here that the problem lies,
for what exactly constitutes a metalinguistic act *in spatial terms*? What
does it mean for a space to adopt (the representation of) another space
as its ordering 'plan' – to adopt a spatial metalanguage? What tools do
we, as geographers, have to recognize and codify such spaces, such uses
of space? There is only one notion, to my mind, that has, thus far, been
able to successfully speak to spaces that allude to other spaces, places
that stand for other places. Here, of course, I am referring to the notion
of 'heterotopia' proposed by Michel Foucault.[2]

For Foucault (1994: 23), space today has taken on "the form of relations among sites": as he notes, "we do not live inside a void inside
which we could place individuals and things . . . we live inside a set of
relations that delineates sites which are irreducible to one another and
absolutely not superimposable on one another." There are certain sites,
however, which "have the curious property of being in relation with all
the other sites, but in such a way as to suspend, neutralise, or invert the
sets of relations that they happen to designate, mirror, or reflect." Such
sites, according to Foucault (1994: 24) belong to two broad categories.
The first of these are *utopias* – "sites with no real place . . . [which]
present society itself in a perfected form, or else society turned upside
down . . . fundamentally unreal places." The second category are what
Foucault (1994: 24) terms *heterotopias*: "there are also, probably in
every culture, in every civilization, real places – places that do exist and
that are formed in the very founding of society which are something like
counter-sites, a kind of effectively enacted utopia in which the real sites,
all the other real sites that can be found in the culture, are simultaneously represented, contested, and inverted. Places of this kind are
outside of all places, even though it may be possible to indicate their
location in reality." Places which stand out clearly and which are
"absolutely different from all the sites they reflect and speak about."

Landscape as Heterotopia

What kinds of places qualify as heterotopias? Foucault lists several and
all seem to be pertinent, at least in part, to the argument I have been
trying to elaborate. First come the theater, the cinema, and the garden.
This last could, perhaps, be seen as the original heterotopia, if we intend
the latter as a microcosm which both symbolizes and embraces the
macrocosm. The inventory continues with the museum, the library, fes-
tivals, and holiday villages, cemeteries etc.; the list outlined by Foucault
is, in fact, quite long. What is missing from it, however, is the notion

of landscape. Yet, I would argue, rethinking landscape in heterotopic terms might prove quite meaningful: for the first time ever, a portion of space could (re)configure itself (or be reconfigured) as a heterotopia *without a corresponding functional place* but, rather, merely in virtue of its *framing* (of space) (such as that of a photograph, of a film shot), with an apparently innocent conceptual shift producing an extraordinary ontological 'mutation.'[3]

Landscape as the Heterotopia of Modernity

A landscape is either envisioned as such or it is not. There are no half measures in this regard. Landscape is a conceptual device, a mental 'filter' which, once placed upon our eyes and minds, cannot be taken off. Once activated, it no longer needs particular signs, particular indications to produce its effects.

We could even accept the above proposition if we conceive of landscape as simply a heterotopia. But if Augustin Berque (1994: 6) is right,[4] landscape is a very *particular* sort of heterotopia – it is a heterotopia unique to the Modern world:

> la moindre investigation historique ou anthropologique révèle immanquablement ce fait, inadmissible pour notre sens commun: *le paysage n'existe pas comme tel à toutes les époques, ni dans tous les groupes sociaux.* En Europe notamment, c'est une notion qui n'est apparue qu'à la Renaissance. Plus renversant encore: la beauté grandiose de la nature sauvage – celle par exemple que nous admirons dans les Alpes – n'a pas été reconnue avant le XVIIIᵉ siècle. Par ailleurs, des grandes civilizations, comme l'Inde ou l'Islam, ont appréhendé et jugé leur environnement dans des termes irréductibles à la notion de paysage. En réalité, le mot paysage, les tableaux de paysage, l'exaltation esthétique et morale du paysage sont des phénomènes particuliers, dont la plupart des cultures ne donnent pas l'exemple.

It is not by chance that landscape as a 'way of seeing' (Cosgrove 1984) was born in Renaissance Europe. The map, Renaissance perspective painting, and the landscape ideal are all closely related – all three the 'effects' of the linear projection of a multidimensional world. Since within these few pages I cannot possibly hope to demonstrate this above effect in any sufficient detail, I will limit myself to highlighting some of its more relevant implications.

Landscape, as the archetypal heterotopia of Modernity, presents a problem of particular relevance in today's world which proclaims its respect for Otherness and difference, and within which we cannot simply claim the right to speak in others' names. We should thus take heed of Augustin Berque's admonition that not all civilizations have codified the relationship between human beings and their environment in 'landscape' terms. As a renowned scholar of Oriental civilizations, Berque (1994) maintains that not only is the landscape ideal alien to a number of cultures (such as numerous Islamic ones), it is a relatively new concept within our own culture as well. While, for one, landscape would first be nominated and described in China already in the fourth century AD, in Europe, the landscape ideal would only take hold from the sixteenth century onwards.[5]

What does this mean? Berque's (1994) analysis circumscribes 'landscape culture' and inscribes it within the perimeters of our own distinct 'semiosphere' (to use Lotman's (1985) terminology).[6] When we relate to the Ancient Greeks or Romans in landscape terms, we impose an idea(l) of landscape upon classical civilization which is, fundamentally, alien to the culture of those times. We should thus certainly question the legitimacy of inscribing an ancient Roman aqueduct or a mosque within a landscape perspective; a question which should also be of first order whenever we interact with any other community or social group not (yet?) 'addicted' to the landscape idea(l).[7]

In this sense, landscape is much more interesting than Foucault's library or museum, maybe even more so than the cinema and the theater, as its 'heterotopic' subversive potential is much greater. Once we begin to see a landscape, we see it everywhere, and the whole world appears to us as but a constellation of landscapes. But if we conceive of it within the limits of our culture, if we frame it within that particular cultural project that we have termed Modernity, then landscape becomes a *threshold* – through it, we enter the Modern project – but perhaps through it, we can also abandon it. But to exit from the confines of Modernity, we must somehow suspend or invert our by now common-sense notion of landscape; a task possible only if we contrast the idea(l) of landscape with another conceptual tool capable of neutralizing the force of this latter. This is why, for quite some time now, I have been exploring the possibility of applying cartographic rhetoric to the landscape idea(l) in order to probe the subversive potential dormant within cartographic instruments.

Landscape and Cartographic Rhetoric

For Foucault, the mediating device between utopia and heterotopia is constituted by the mirror – and here, we cannot but turn to the work of Massimo Quaini (1994: 321) who has long argued for a history of cartography as a 'specular' narrative:

The task of the historian of cartography is to codify the rules, to order the endless succession of diverse maps through time, deposited in a sort of cartographical stratification and awaiting the patient expertise of an able archaeologist; a succession which often appears, however, something of a Baroque hall of mirrors, designed to disorient, to bewilder anyone who enters its doors. The opposite effect, in fact, of that proclaimed as the role of the single map which, at least in theory, is meant as an orientation device . . . The history of cartography, deriving as it does from the history of vision and visual logic, has thus also taken the form of a labyrinth of mirrors.

The historian of cartography, by focusing not on the single map but on a whole *set* of maps, is thus able to 'uncover' some of the underlying cartographic rhetorical devices. We can note here one such critical 'passage' which characterized the modern era: that is, the shift from map collections to Atlases. As Giulia de Spuches (1996: 44) notes:

One of the most innovative aspects which can be traced to the emergence and use of Atlases (as ordered and indexed collections of cartographical representations), is the development of one particular property of the map: that is, nonlinearity, which the user can now employ by leafing through its pages in one direction or another, now able to shift scales in continuous fashion. Scale, in fact, is no longer uniform; the reader can thus follow his or her own trail, with the limitations imposed by the cartographic medium no longer ossified within a rigid hierarchy. S(he) can now negotiate between the global and the precision of a point in space; can 'read' the map in several directions. The wall chart, the singular map, the map projected from a singular perspective, asserted the gaze of the strong subject, a gaze which erased the relative, transcalar properties of space. The Atlas, with its scalar flexibility, on the other hand, is able to embrace the abundance of toponyms, the multiplicity of detail and diversity of points of view, albeit always within the confines of the order which frames its construction. The Atlas is, therefore, always the expression of a unified intellectual project – an *ante litteram* instrument of the emergent forms of encyclopedic knowledge (Jacob 1992). Cartographic logic, which had already succeeded in reducing the variety of geographical 'contents' to a geometrical ordering, would find in the Atlas a brave new tool in the organization of the rhetoric of representation.

The Paradoxical Game of the Modern and the Postmodern

The renewed attention to metaphor in geographical discourse (Dematteis 1985) (as well as within social-scientific thought more broadly) may grant some very useful insights here, allowing us to draw upon the many possibilities offered by the field of *rhetoric*.

The interrogative rationality of rhetoric (see Meyer 1997) expresses itself within *tropes*, or figures of style. We can define a trope as the shift in a word or a phrase from its 'proper' meaning to another. In a certain sense, key tropes act as the markers of identity, 'applied' in differential fashion (Meyer 1997: 130). And it is within the trope that we find the elements of that which Anglo-Saxon scholars have termed the "willing suspension of disbelief," that voluntary suspension of belief which allows us to accept as real that which is nothing else but pure fiction (Meyer 1997: 115). Within tropes, we can locate the trail of linguistic innovation – where language still oscillates between the nominative and the subject, where it still interrogates rather than emitting final judgment, where it still contemplates before answering (Meyer 1997: 115).

But to allow for such innovation, even rhetorical figures must operate a division within themselves; they must oscillate, must enter into a paradoxical game with themselves. It is only thus that they are able to intentionally 'miss the mark,' to cross the threshold:

> Their apparent conformity – to the letter – allows them to miss the mark; the only way in which they are able to conserve their effect, once enunciated, is by rejecting translation, by playfully reconfiguring themselves in order to delay translation. It is a question of time – of gaining time. Of taking advantage of that tiny break which constitutes the opening – and of delaying as much as possible the eventual closure of the model. (Rovatti 1997: 58)

Michel Meyer's (1997: 23) definition of rhetoric is quite relevant for geographers: "it is the act of the negotiation of the distance between individuals with respect to a distinct dilemma." In this sense, a communicative process which draws merely upon the *literal* sense of words is but an extreme – usually the final result of a much broader universe of discursive practices which incessantly meld and transfigure literal and figurative meanings. Within this linguistic game between the interlocutors, something very important occurs: for not only do the actors bring into focus the object of discourse, observing it from diverse and distinct points of view – they also assume this latter as the privileged reference point around which to negotiate the distance which separates them one from another.

Of particular interest is also Kenneth Burke's (1969) formulation of a system of symmetries between rhetorical figures and the devices which give birth to the cartographic image. Burke in fact associates metaphor with cartographic projection, metonymy with the concept of reduction,

METAPHOR	PROJECTION
METONYMY	REDUCTION
SYNECDOCHE	REPRESENTATION
IRONY	DIALECTICS

Figure 10.1 From Burke, 1969

synecdoche with that of representation, while irony is granted a dialectical function (see Meyer 1997: 112, and figure 10.1).

The Willing Suspension of Belief

It is only irony, then, which has not been transposed into cartographic terms. This should not surprise us, however, for irony commands a very particular rhetorical status – as does cartography within geographical discourse. Cartography, in fact, asserts its authority as the image of the world *precisely by virtue of its nature as a language free of irony, exempt from dialectic*. The admission of these latter would be much too disruptive for cartographic knowledge. But irony is the best suited, from among all the rhetorical figures, to set in motion (and, perhaps, conclude) this game of endless paradox, precisely due to some of its very particular properties. First of all, it is the only rhetorical figure which is not of an essentially linguistic nature; after all, none of the other rhetorical figures can be taken literally. But in the case of irony, it is the *context* alone which allows us to decide whether the speaker has just asserted one thing – or its polar opposite – since the statement itself does not contain any conceptual-linguistic contradictions (Meyer 1997). It is precisely this contextualized nature of irony which allows it to 'reveal' subjectivity by highlighting the distance between interlocutors (something which does not occur with the other rhetorical figures), thus allowing it to unmask the play of rhetoric, understood as the *negotiation of distance between subjects*. And once the interrogative rationality of the rhetorical field has been made explicit, the artifice is exposed and irony grants us access to a second discursive 'level.' It is precisely this ability which allows irony to overcome the stage of simple verbalization, manifesting itself, rather, as a subjective attitude which, consequently, reflects the distance between subjects (Meyer 1997: 127–9).

Of course, the subversive effects of irony would prove entirely too virulent for cartography: that distinct discursive universe within which the *ad rem* and *ad hominem* levels of rhetorical argumentation coincide. Here, the act of the negotiation of distance between subjects (within which all rhetoric is based) is manifested precisely within . . . the negotiation of distance between subjects. And since there is no margin, no break separating the *ad rem* from the *ad hominem*, the rhetorical shift (which usually occurs when an *ad rem* argument is particularly problematic) cannot occur. If rhetorical figures consent the 'willing suspension of disbelief' – which, as I have noted previously, allows us to accept as real that which is nothing else but pure fiction (Meyer 1997) – in cartography, irony can perhaps allow for the opposite effect: a sort of voluntary 'suspension of belief.'

The rhetoric of cartographic representation, now rendered even 'stronger' by the adaptability and malleability of digital tools such as geographical information systems, has greatly extended its boundaries, though it remains subordinate to the disorientation produced by that illusionary game of mirror-image which I described at the outset. But what would occur if we chose to, willingly, apply this labyrinth of mirrors to landscape? Would not an opposite effect occur – precisely because it would find itself confronted with the very same device? GIS, as the 'digital representation of the landscape of a place' (as Dobson 1993 terms it[8]), by envisioning landscape as simply the 'space of analysis' could, therefore, prove an essential tool in suspending and neutralizing its effects. Read through the (digital) cartographic lens, even landscape, then, reveals its true nature as a specular image of the world. And so the history of Modernity unfolds before the gaze of those able to understand that which is recounted by the history of cartography, that which is revealed by the European landscape.

The Threshold of Modernity

To gain access to places, we must first subvert landscapes. If perspective is nothing other than the 'symbolic form' of the emergence of the Modern Subject, we can view the discovery of landscape as, accordingly, the symbolic form of the emergence of the Modern world, objectified by the gaze of this new subject.

But we can breach the threshold which, since the onset of Modernity, has separated the Subject from the World precisely in virtue of the fact that landscape, just as every other heterotopia, both opens and closes with respect to other places. Geographical information systems, as the *ratio extrema* of cartographic logic, furnish a digital representation *of* the landscapes *of* places – and so render explicit the distance between the modern subject and the world. By revealing the rhetorical devices of landscape and thus 'suspending belief,' they can allow, I would claim, for a *'post'*-Modern reading of space.

A willingness to question the meaning of landscape today allows us to place ourselves on the threshold of Modernity; allows us to place ourselves in a critical position from which we can operate a fundamental choice: whether to continue within the confines of the Modern project, or have the courage to step outside of them. (Only to subsequently claim, perhaps, that we have never been Modern, as Bruno Latour (1995) would have it.)

NOTES

1   Perhaps the Tower of Babel should be reimagined not as a chaos of a multiplicity of languages but, rather, as a stratification of metalanguages.

2   For an overview, see the 1994 issue of *millepiani*, or its geographical interpretation provided by Soja (1996).

3   A shift which may seem contrived on the spot but which, as we shall see, has a long and complex cultural history.

4   A line of thought also reflected in the work of Franco Farinelli – see, in particular, his essay on the "Witticism of Landscape" (1992) as well as his *Geotema* (1995) article.

5   "Quant à moi, j'ai empiriquement adopté le quatre critères suivants, pour distinguer les civilisations paysagères de celles qui ne le sont pas: 1. usage d'un ou plusieurs mots pour dire 'paysage'; 2. une littérature (orale ou écrite) décrivant des paysages ou chantant leur beauté; 3. des représentations picturales de paysages; 4. des jardins d'agrément. C'est le premier de ces critères qui est le plus discriminant, et l'histoire montre qu'effectivement il implique les trois autres. De très nombreuses cultures ne présentent ou n'ont présenté aucun des quatres critères. Le grandes civilisations ont toutes présenté au moins l'un des trois derniers. Seules deux d'entre elles, dans l'histoire de l'humanité, ont présenté l'ensemble des quatre critères et notamment le premier: la Chine à partir du IVe siècle de notre ère, et, mille deux cents ans plus tard, l'Europe à partir du XVIe siècle" (Berque 1994: 16).

6   On the relevance of this notion for cultural geography, see my (1996) essay.

7   "La question, pour nous qui baignons dans une civilisation paysagère, c'est d'arriver à comprendre, ou ne serait-ce qu'à admettre, que d'innombrables cultures, et plusieurs grandes civilisations, ont eu conscience de leur environnement dans des termes qui sont irréductibles au paysage. Des termes que nous ignorons tout autant qu'elles ignorent la notion de paysage, voire ignorent l'ensemble des quatre critères définis ci-dessus. Leurs critères à elles, nous y sommes tout aussi aveugles, et nous n'avons pas de mots pour

les dire; à moins d'un patient, d'un humble travail d'apprentissage et de traduction . . ." (Berque 1994: 16).

8 "I would define GIS as a digital representation of the landscape of a place (site, region, or planet), structured to support analysis" (Dobson 1993: 434).

REFERENCES

Berque, A., ed. (1994). *Cinq propositions pour une theorie du paysage*. Paris: Champ Vallon.

Burke, K. (1969). *A Grammar of Motives*. Berkeley: University of California Press.

Cosgrove, D. (1984). *Social Formation and Symbolic Landscape*. London: Croom Helm.

Dallenbach, L. (1994). *Il racconto speculare. Saggio sulla mise en abyme*. Parma: Pratiche.

Dematteis, G. (1985). *Le metafore della Terra*. Milan: Feltrinelli.

Dematteis, G. (1995). *Progetto implicito. Il contributo della geografia umana alle scienze del territorio*. Milan: Franco Angeli.

De Spuches, G. (1996). Atlanti e ipertesti. *Geotema* 2(6): 640–5.

Dobson, J. E. (1993). The Geographic Revolution: A Retrospective on the Age of Automated Geography. *Professional Geographer* 45(4): 431–9.

Duncan, J. S. (1990). *The City as Text: The Politics of Landscape Interpretation in the Kandyan Kingdom*. Cambridge: Cambridge University Press.

Duncan, J. S. (1992). Re-presenting the Landscape: Problems of Reading the Intertextual. In *Paysage et crise de la visibilité*, eds. L. Mondada, F. Panese, and O. Soderstrom. Université de Lausanne: Institut de Géographie, 81–91.

Farinelli, F. (1992). *I segni del mondo. Immagine cartografica e discorso geografico in età moderna*. Scandicci: La Nuova Italia.

Farinelli, F. (1995). L'arte della geografia. *Geotema* 1(1): 139–55.

Foucault, M. (1994). Eterotopia. Luoghi e non luoghi metropolitani. *millepiani* 1(2): 11–20.

Gambino, R. (1989). I piani paesistici nell'esperienza urbanistica. *Rivista Geografica Italiana* 96: 427–43.

Gambino, R. (1991). *I parchi naturali. Problemi ed esperienze di pianificazione nel contesto ambientale*. Rome: La Nuova Italia Scientifica.

Guarrasi, V. (1996). I dispositivi della complessità. Metalinguaggio e traduzione nella costruzione della città. *Geotema* 2(4): 137–50.

Jacob, C. (1992). *L'empire des cartes. Approche théorique de la cartographie à travers l'histoire*. Paris: Editions Albin Michel.

Latour, B. (1995). *Non siamo mai stati moderni. Saggio di antropologia simmetrica*. Milan: Eléuthera.

Lotman, J. M. (1985). *La semiosfera. L'asimmetria e il dialogo nelle strutture pensanti*. Venice: Marsilio.

Meyer, M. (1997). *La retorica*. Bologna: Il Mulino.

Quaini, M. (1994). La carta geografica. Un racconto speculare. *Rivista Geografica Italiana* 101: 319–26.
Rovatti, P. A. (1997). La carriola e la segatura. *Aut Aut* (282): 53–60.
Soja, E. W. (1996). *Thirdspace: Journeys to Los Angeles and Other Real and Imagined Places*. Oxford: Blackwell.
Zerbi, M. C. (1993). *Paesaggi della geografia*. Turin: Giappichelli.

11

Mapping the Global, or the Metaquantum Economics of Myth

Franco Farinelli

The Map, the Quantum, and the Critique of Cartographic Reason

The critique of cartographic reason elaborated to this day has taken on two distinct forms. The first explicitly interprets Kant's *Critique of Pure Reason* as the protocol of the cartographic *diktat* (Farinelli 1995; 1997a; 1998a; Olsson 1998). The second strives to reconstruct the means of control inherent to cartographic reason, focusing in particular on the uses of logic, language, and images of the world (Farinelli 1976; 1981a; 1981b; 1989a; 1989b; 1992). Both critiques, however, herald the demise of classical geography.

Certainly, the notion of 'classical geography' can be interpreted in many different ways. For Richard Hartshorne (1961: 35–83), the 'classical period' in geography was that marked by *Erdkunde* in the first half of the nineteenth century – the sole instance in modern geography when the "cartographic dictatorship" (Ritter 1852: 34–5) framing geographical thought would come under fire with the emergence of critiques against all those who "erroneously believe we deduce from the external world that which, instead, we ourselves have imposed upon it" (Humboldt 1845: 8). According to Paul Claval (1971), on the other hand, classical geography evolved as the domain of the French school, ending at the beginning of the second half of the twentieth century.

My understanding of classical geography is quite different from Hartshorne's. Within this chapter, classical geography is conceptualized, rather, in terms similar to those defining classical physics: a geography within which the relationship between the object of analysis (the surface of the earth) and the measurement apparatus (the map) – and thus also the research subject – is not considered worthy of attention and excluded from analysis. Quantum physics, however, has argued since the beginning of this century that this relationship was, in fact, inseparable from the phenomenon itself (Bohr 1965: 103–210). It was Max Planck, the 'father' of quantum mechanics, who would rediscover *Erdkunde*'s lesson half a century after its defeat within geography: that is, that not only is the observer inseparable from that which (s)he observes but that, rather, (s)he acts in part to determine it (Farinelli 1998b).

The critique of cartographic reason is based, above all, on the recognition that the secret to all of Western knowledge lies in projection, a secret that Kant himself had discovered – and by silencing it, rendered it absolute. "*De nobis ipsis silemus*" he would note: let us not speak of ourselves. This is the motto (taken from Bacon) with which Kant opens his first critique – in this fashion, silencing the fact of being a geographer (Farinelli 1999a). Gregory Bateson was fond of saying that a theory is nothing other than the description of a description. Kant's proposed solution to the problem of knowledge was, therefore, simply the description of Ptolemy's description of how to transform the globe into a map. But Kant's teaching would, at the same time, adopt Ptolemaic thought to transform that which is subjective into something that is objective. Objectivity is nothing other but the result of a transcendental transformation of subjectivity. It is thus that the point of projection becomes the subject – and cartographic representation takes the place of the world; a world which is knowable in 'objective' fashion *only* through such a substitution (Farinelli 1995: 143–6). What emerges is an insurmountable chasm between subject and object, and a perfect correspondence between this latter and the tools adopted to describe it. It is precisely this insurmountable chasm – which is, at base, of an ontological nature – and the conflation between the world and its image (to use Heidegger's expression) that has characterized most of classical or modern geography; a geography which, apart from the brief period of critique articulated by German geographers at the beginning of the nineteenth century, has essentially ignored the relationship between subject and object (mediated, as it was, by the cartographic apparatus). It is for this reason that postmodern geography, emergent from the crisis of modern geography, is assuming (as yet unawares, perhaps) quantum forms.

It may well be true, as Kevin Kelly (1995: 25–6) maintains, that the atom is too simplistic, that it is passé, and that the future of knowledge lies with a "dynamical net" able to channel "the messy power of complexity." Within geography, however, the evolution of epistemological reflection lies far behind that of atomic physics, and even recent conceptualizations of network logics could well learn from the theorizations of the latter. Manuel Castells (1996: 3) has noted, in fact, that our societies are increasingly structured around a bipolar opposition between the net and the self, in a condition of structural schizophrenia between function and meaning. However, how do we transcend this problem? One possible solution (once having traveled through the territory of quantum physics) is offered by none other than the rediscovery of that ancient and familiar land of myth. We can thus begin to understand the process of globalization only if we conceptualize it as but an extension of the processes of quantum physics into the 'physics' of the world at large – and if we assume that it is only in myth that we can find models able to explain the logics behind the workings of the world. It is thus myth that gives form to the quantum nature of the world. Our world is a quantum world, and myth is its logic.

It is not by chance, then, that apart from vague references made to it in the past by Peter Haggett (1965: 25) and R. A. Briggs (1987), quantum theory has entered geographic discourse only of late and, in particular, only after the fall of the Berlin Wall (Harrison and Dunham 1998; Collier et al. 1999; Spedding 1999; Harrison and Dunham 1999). It was the fall of the Berlin Wall which signalled the final end of the project of the territorial production of modernity – a project *not* founded upon cartographic representation as a reflection of the territory but, rather, upon the material construction of territory as a reflection of cartographic logic. It was the Wall, more than any other construction, that came to represent the materialization of an abstract line – made visible, legitimized and authorized precisely by cartographic representation (Farinelli 1999b). Jean Baudrillard (1981: 10) is wrong, then, when he argues that postmodernity is characterized by the fact that the map now precedes territory. Quite the contrary; in *all* of modernity (including classical physics) the map has had precedence and, I would argue, modernity itself consists in nothing other than just such a structural ordering (Farinelli 1989b). In fact, it could be argued that the precedence granted the map underlies the very bases of Western thought more broadly (Farinelli 1998b). To contradict Baudrillard, I would suggest that postmodernity emerges precisely from a crisis of this precedence of the simulacrum. In postmodernity, the world no longer functions as a map – as a reduction to 'space,' to a measure of distance. The world can thus no longer be represented within what I term 'tabular logic' – above all, within the Euclidean attributes of extension of continuity, homogeneity, and isotropy (Farinelli 1983: 27).

Quantum mechanics forced physics, before any other form of knowledge, to face (albeit unawares) the crisis of the validity of the cartographic model and of its implicit assumptions – and thus also of the very notion of an objective reality (and of the univocal nature of the relation between name and thing, between signifier and signified) which could spring only from such a model. Equally unawares, geography would arrive at similar conclusions much later. It was only at the end of the 1960s that Anglo-American and European geographers (albeit in diverse ways and largely unbeknownst to each other) began to note that widely diverse if not contrasting historical and social processes often materialized in identical fashion – material manifestations such as the dwelling types analyzed by the descriptive settlement geography inspired by the French tradition (Pecora 1970), or more 'immaterial' manifestations such as the curves of the Cartesian axes which formed the object of study of quantitative geography (Olsson 1991: 53).

In their 1998 *Transactions* paper, Harrison and Dunham (1998: 508) cite Bohm's reflections about classical physics as the construction of a "fragmentary view of the natural world which is therefore reflected in our use of language to categorise and subdivide elements of this world." Yet this fragmentary view is, above all, the product of the reduction of the world to a map; it is the result of a *mapping* (Farinelli 1990) within a scheme imposed by Cartesian axes. Every map is thus but a simple Cartesian space: the only space within which points are objects which can be precisely located; the only space within which the metric 'works' since linear distances have a meaning. But within Hilbert's space of quantum mechanics characterized by infinite dimensions, bidimensional Cartesian space functions as merely the lower limit and the metric makes no sense. What then?

There are two paradoxes which should be noted here. The first is Heisenberg's (1961: 64–5) observation that the Kantian *a priori* is indirectly grounded in experience, envisioned as deriving from an ancestral formation of the human mind – quite similar, in fact, to Lorenz's stipulation (for the animal world) of the existence of distinct hereditary forms or innate patterns of behavior. What Kant did not foresee, however, is that for notions such as space, time, and causality, *a priori* synthetic understandings could furnish only relative truths. For Heisenberg, in fact, this fundamental paradox of quantum theory lay outside of the bounds of Kantian logic. The second paradox emerges from recent reflection in quantum mechanics. Despite this latter's assumption of the lack of an objective reality, it nevertheless stipulates the existence of a macroscopic world which, for the most part, acts *as though it were* objectively real. We are, therefore, led to believe in an objective reality since quantum mechanics predicts that the world should function according to such 'objective' rules. But, on the other hand, it is precisely *because* the world functions in a certain fashion that we harbor such a deep faith in the existence of an objective reality – and it is this conflation which renders understandings of quantum mechanics so problematic (Lindley 1997: 248).

These two paradoxes can, however, both be resolved by means of a single assumption: that in the construction of Western knowledge, it is the map which corresponds to that which in quantum theory is referred to as the decay or the reduction of the wave function – a reduction that is, fundamentally, the result of what geographers term projection. Yet although geographers may know everything there is to know about the rules of projection, I would argue that they entirely ignore its deeper nature. Following Tissot's theorem, we know that to reduce a sphere to mappable form, we must choose between a precise knowledge of the form and a precise knowledge of the area – just as Heisenberg's principle necessitates a choice between the knowledge of the position and the knowledge of the speed of a particle inside the quantum field (thus, of any object within any delimited spatiotemporal field). Just like a wave function, a projection contains a synthesis of the information necessary to calculate the probabilities of the different results of a given measurement. And, just like a wave function, a projection is a description of a system – that is, both of the object measured, as well as of its measurement. This signifies that a projection does not simply depend on the object in question, but also on the measurement chosen – and its intended uses.

The decay or 'collapse' of a projection – that is, the map – ensures the elimination of 'de-coherence' (that is, the elimination of the probability of entanglement, of a lack of coherence). Yet it is also a measurement device which, by virtue of its reduction to scale, already delimits all possible measurements/metrics of an object or system – precisely due to the extreme limitation that it imposes on the possibilities of measurement itself. It is the map, the macroscopic device *par excellence*, that transforms a state of uncertainty into certainty, that renders that which is open and undefined as determined and finite, thus furnishing the bases for that which we conceive as a classical, deterministic system. It is precisely cartographic representation which acts as the agent of the process that W. H. Zurek (1993; 1994) terms "environment-induced superselection," by which "only a small subset of possible states is chosen for a stable existence." Quantum mechanics thus points us to a critique of cartographic reason located within the broader scientific debate. At the same time, however, elaborations of this critique may, in turn, shed new light on some of the classical paradoxes of quantum mechanics as well.

The Globe, the Net, the Myth

Manuel Castells' (1996: 23–4) recent work has focused on defining the nature and the rules governing 'informational capitalism.' Castells grounds his analysis within one of the paradoxes highlighted by Raymond Barglow (1994): namely, that while information systems and networking augment human powers of organization and integration, they simultaneously subvert the traditional Western concept of a separate, independent subject. As Barglow (1994: 6) himself has stressed, "the historical shift from mechanical to information technologies helps to subvert the notions of sovereignty and self-sufficiency that have provided an ideological anchoring for individual identity since Greek philosophers elaborated the concept more than two millennia ago." To my mind, this signifies that subjectivity – as objectivity's symmetrical counterpart – is also a product of the "cartographic ethos" (Farinelli 1997b: 43–5; Conley 1997) and that its present crisis is similarly linked to the crisis of cartographic models of the world.

If the above is true, it reveals a fundamental contradiction within Castells' work, as well as its principal limitation. On the one hand, Castells (1996: 61–2) attempts to grasp the logic of the new techno-economic paradigm as, essentially, a network logic which reflects the topological nature and the unpredictable character of its development patterns. On the other, however, the variable geometries that he traces and that, according to him, are fast dissolving existing historical and economic geographies (Castells 1996: 106–99) are anything but differential, for they limit themselves to simple *internal* variations within an all-too-familiar Cartesian space (within which all coordinates are necessarily endowed with a defined metric) (Auyang 1995: 26–30). In like fashion, Castells (1996: 65) recalls Kranzberg's first law which states that "technology is neither good, nor bad, nor is it neutral" – but fails to recognize that this principle operates within the logic of quantum mechanics, "which has a *greater logical coherence* than classical mechanics" (Birkoff and von Neumann 1936: 836–7) and which postulates that all experimental questions can only be answered in terms of *probability values*. Unlike (Aristotelian) cartographic logic, the logic of quantum mechanics is not limited to binary solutions; within it, *tertium datur*. Just like Kranzberg's law, it follows the fundamental quantum principle of complementarity (Dalla Chiara and Giuntini 1997: 615; Heisenberg 1961: 181–3; Bohr 1965: 104–8). In fact, if we pay close attention, the world described by Castells appears to function precisely according to the basic principles of quantum theory – and we can only understand this world if we adopt the basic principles of quantum logic.

According to Castells, the informational economy is global. Unlike the modern world economy and its diffusion of capital accumulation throughout the world, the informational economy is able to function as a *unit* in real time on a planetary scale (Castells 1996: 92). But this new economy is not a planetary economy proper, for its structure comprises only fragments and portions of countries and economic regions, thus

forming a system that is, all the while, "highly dynamic, highly exclusionary, and highly unstable in its boundaries" (Castells 1996: 102). The
informational economy is thus "deeply asymmetric. But not in the simplistic form of a centre, semi-periphery and a periphery, or following an
outright opposition between North and South; because there are several
'centres' and several 'peripheries,' and because both North and South
are so internally diversified as to make little analytical sense in using
these categories" (Castells 1996: 108). This passage speaks directly to
the crisis of the cartographic (that is, Euclidean) model of the world
that, following Ptolemy, would become the dominant geographical
model. It was Ptolemy, after all, who privileged the map over the globe,
arguing that the globe was extremely awkward for it forced its user
either to move *around* it – or to move *it* around by hand. The map, on
the other hand, rendered either action unnecessary. Although Ptolemy's
recommendation may appear to be of a practical nature, it is of great
importance for it points to the very nature of modernity itself – which,
I would argue, is essentially Ptolemaic.

Already during the flowering of the Renaissance, Giorgio Vasari
insisted that statues be crafted in a fashion that would permit their
viewing from 16 different perspectives – thus requiring a much more
elaborate workmanship than classical Greek statues, which were
designed with but four distinct perspectives in mind (Wittkower 1985:
14). The globe, however, the very prototype of baroque sculpture, was
a work of art conceived with a single viewer in mind; a viewer who
would admire this representation by circling around it. Yet if the globe
is a statue of the Earth, *where* does its viewer move about? (S)he is, necessarily, external, traveling around the delimited and finite globe in what
could only be an infinite and empty space (Schmitt 1981: 58).

How was such an infinite emptiness conceivable before the first astro-
naut – in a sense, the first truly postmodern subject since the first able
to evade the modern correspondence between the world and its cartographic image? Still, in Voltaire's day, the vision of our planet surrounded by an infinite void both horrified and fascinated the Parisian
salons (just as did the existentialists' *néant* in more recent times). But
the action of circling around the globe (so highly discouraged by
Ptolemy) does not merely contribute to the implicit admission of the
existence of an absolute empty space. It also allows for the elaboration
of a way of knowing founded upon the admission that knowledge itself
is temporally determined: both as a process, but also as a historically
contextualized form of learning. Rotating the globe itself while standing still, on the other hand, implies a way of knowing founded upon
both sight and touch. What is important, however, is that *both* cases
(both of which run counter to Ptolemy's prescriptions) contradict the

fundamental principles of modern epistemology; principles which have closely delimited the relationship between the viewing subject and cartographic representation in a manner unchanged since the advent of perspective vision at the beginning of the fifteenth century. The modern subject is still "as though paralyzed by *curare*" (as Florenskji (1984: 83) would note ironically) and knowledge itself is simply the product of the gaze, a gaze able to implicitly and instantaneously (and thus a-temporally) erase the distance which separates it from the object of observation. Yet by adopting the globe, it is impossible to separate time and space in modern fashion. The globe renders Kantian logic impossible – and perhaps this is why Nietzsche insisted on a 'dancing thinking' and thus one that was, implicitly, global.

Paradoxically enough, for a subject that travels around the sphere of the globe, space does not exist – or at least the Ptolemaic vision of space as an abstract standard metric interval between two points. On the globe, there is no scale, since the proportions of its parts are determined, rather, in terms of internal reciprocal relationships and thus according to a self-referential logic. We should also recall that in medieval times, it was the globe which represented God: as a sphere whose center is everywhere and whose circumference is nowhere. In the seventeenth century, considered by many the golden age of the globe (van der Krogt 1993), it was in just such terms that Pascal would describe Nature itself. Yet Pascal's understanding, considered labyrinthine, would be rejected by Western knowledge as even more dangerous than the possibility of the existence of a void (Farinelli 1997b: 39). In the simplest of terms, the sphere and the plane – the globe and the map – differ topologically since the holistic properties of their surfaces are diverse. One is thus irreducible to the other since the sphere is rotund and finite, while the plane, on the other hand, is open and its straight lines unbounded (Reichenbach 1957: 59). It is precisely this characteristic of that which Lefebvre (1978: 59) has termed the "space of representation" that has allowed for the uninterrupted expansion – both economic as well as cognitive – of Western modernity. The "work of art" that is the modern state (as Burckhardt 1980: 5–124 terms it) has thus always been conceived within the limits of Euclidean logic – by privileging extension at the expense of global models and thus privileging territorial continuity, cultural homogeneity, and an isotropism reliant upon the existence of a sole center, that of the state capital.

But within the logic of the sphere or globe which does not admit the existence of straight lines, a minute surface area may correspond to an infinitely larger volume (Volk 1995: 10–13). It is a logic which closely mirrors that of capitalist accumulation which is highly selective and thus, necessarily, discontinuous; fragmentary and thus nonhomogenous;

anisotropic since differential and, above all, characterized by the absence of a singular center; founded, in fact, upon the very existence of a plurality of possible virtual centers – imaginable only upon the surface of a globe. This is also the logic that Hegel describes in his dialectical account of history, founded upon a distinction between that which is necessary in rational terms, and that which simply constitutes a residual; a conceptualization driven by a principle of selectivity grounded within things themselves, the only principle able to define the identity of the real and the rational. Hegel's vision, of course, preceded the triumph of positivist thought and the formulation of historical models that were, necessarily, cumulative, linear, and inseparable from the cartographic perspective, as well as predicated upon an absolute continuity of causality. But, as Edmund Husserl (1961: 65) has argued, this systematic geometrical (that is, cartographic) reduction of the world already presupposed within its very logic a systematic relationship between cause and effect. Not only; as Ernst Troeltsch (1989: 46) would note, positivist models of historical process ignored the role of flux and the growth of surplus value – thus ignoring the fundamental logic of capitalist accumulation. It is just this logic, I would argue, that Castells has been trying to describe within his work. We should also recall, however, that it is this very same logic that Ptolemy was unable to describe. But Ptolemy was a subject of Imperial Rome at the height of its power – and no Empire has ever tolerated polycentric representations of the world. Thus the Ptolemaic propensity for monocentric (and thus cartographic) representations of the world. The persistence of this cartographic ethos in all fields save physics has assured that modernity has remained, for all intents and purposes, essentially Ptolemaic, centuries after the latter had formulated his first astronomical models. In his attempts to narrate the global dimension of today's world, Castells too finds it difficult to transcend the inescapably cartographic nature of his analytical models. But how else is it possible to conceptualize an informational capitalism which functions in real time on a planetary scale? How does Castells' 'space of flows' function, then? Ten years ago, at the time of the fall of the Berlin Wall, Castells (1989) traced a dichotomous distinction between places and so-called 'spaces of flows': the former conceived as the site of social reproduction, the latter, as an overlapping 'layer' within which the abstract decision-making of finance capital operated – often by negating precisely the local specificities of places, guided by the invisible codes of international capitalism. Within this early conception, Castells thus stipulated the existence of two distinct dimensions – the one 'physical' and the other, in a sense, 'metaphysical' (albeit able to exert quite concrete and pervasive effects).

The author's position appears to have undergone a significant shift in his latest work, however. The structural divide that Castells traces is no longer one between the material and immaterial realms, but rather one distinguishing contiguity from its absence. The notion of space in its original Ptolemaic metric understanding is no longer present. 'Space' now comes to signify any given set of flows or places, and place "becomes a locale whose form, function and meaning are self-contained within the boundaries of physical contiguity" (Castells 1996: 423). It is worth noting, though, that it is the modern territorial state, as the archetypal expression of Euclidean space, that would have best fit the above definition; certainly a signal of just how far the categories of modernity have been subverted in the 'information age.' Within the space of flows, on the other hand, an action in one region has immediate effects on what occurs in another (counter to Einstein's arguments in his debate with Max Born (Ghirardi 1997: 455)). The new economy could thus be argued to function along the quantum principle of the 'nonlocality' of processes. By this I do not intend merely the notion of 'action-at-a-distance,' for this latter still presumes the transmission of information between sites as well as the presence of observable physical effects. The quantum principle of nonlocality also stipulates, in fact, that which has been termed "passion-at-a-distance" (Shimony 1993: 133; Popescu and Rohrlich 1997) and which does not reside within such physical 'proofs.'

Having reached this point, however, quantum theory ceases to be of use, and necessarily must give place to myth. The philosophy of quantum mechanics is, at this level, still cartographic in a sense, still Cartesian (as Kurt Hubner 1990: 4 argues), for it continues to trace a distinction between subject and object and between the conceptual and material realms. Yet the informational economy does not admit such distinctions; its very specificity, rather, is founded upon their erasure. As Castells (1996: 198) himself notes, "for the first time in history, the basic unit of economic organization is not a subject, be it individual (the entrepreneur) or collective (the capitalist class, the corporation or the State)." The unit, rather, "is the network, made up of a variety of subjects and organizations" – and thus made up both of subjects and objects, of material and conceptual dimensions. These latter categories are by now interwoven to such a degree that they can only be differentiated in purely functional – but no longer ontological – terms. The space of flows, in fact, is not placeless, although its structural logic may be (Castells 1996: 413).

The ontology of myth, however, admits such uncertainty in the nature of the person or thing represented (Nebel 1961: 173). The ontology of myth precedes all dissociation; its way of knowing precedes any category. The world of myth is thus one where everything comes together, where the mind is able to interface directly with reality (Gusdorf 1984: 64–5) prior to cartographic mediation (and it is within this latter, after all, that the triumph of modern spatiality lies). But today, modern spatialities are no longer able to explain much of anything and the ontology of cartographic mediation and its associated interpretative models are increasingly less effective.

It is precisely the pre-cartographic nature of mythical ways of knowing that renders them so precious in formulating models and understandings able to tackle today's world. In a sense, the divide between cartographic and mythical ways of knowing closely reflects the distinction traced by Lévi-Strauss (1962: 26–33) between the engineer and the *bricoleur*. The engineer is, in fact, the archetypal representative of cartographic knowledge, with all of her/his creations necessarily reliant upon: i) the mapping out of a preliminary project or model, as well as ii) the availability of *ad hoc* materials. The act of construction itself is but the third and final phase of the process. The *bricoleur*, on the other hand, relies upon 'savage' (and thus mythical) thinking; her/his work does not necessitate an *a priori* cartographic simulacrum. The *bricoleur's* mythical consciousness situates itself *prior* to any binary divides, prior to the partitioning of the world into mappings – mappings which may render it much more manageable but which, at the same time, *fix it* within a stable constellation of signs. Within myth, however, these latter are always in flux (Lévi-Strauss 1930), despite the fact that myth *does* rely upon totalizing narratives, as Blumenberg (1991: 90) has noted writing on the romantic tradition. For the (be)holder of a myth, reality can only be conceived in 'global' terms; within myth, that which is nature and that which is not are inextricably entwined (Gusdorf 1984: 78).

I would like to add one final observation on the nature of mythical thinking that, I believe, will further highlight its usefulness in coming to terms with the workings of the new informational economy. It is only in myth that we can conceive of a world not composed of a seriality of separate individuals, each with his/her distinct place and thus exclusive one of another (*aut-aut*), and definable only by enumeration (*and-and*); the world of myth, rather, admits the existence of subjects who co-participate, subjects who are mutually interpenetrating (*in*) (van der Leeuw 1945: 45; Cassirer 1964: 75). Michel Maffesoli's (1985: 99–120) notion of 'organic order,' of a differentiated organic nature, relies on a quite similar conceptualization; it is this sort of order, Maffesoli argues, that underlies the 'orgiastic unions' which bind a variety of organizations (such as the Mafia) operating within the informal economy.

If we admit that any understanding of the nature of the new informational economy is necessarily predicated upon the abandonment of cartographic ontology and an adoption of/return to mythical ways of knowing, then the quantum notion of 'passion-at-a-distance' becomes also plausible. The classics of mythological analysis can thus furnish models of/for the informational economy. It is here, I would argue, that the answers to the questions posed by Niels Bohr at the 1927 Solvay Congress lie: namely, why should we abandon, full-scale, the concepts of localization in space and time, as well as the notions of trajectory and individuality (Chevalley 1984: 48)? We should not forget, however, that within the realm of myth, there exists no chronology but merely sequences (Blumenberg 1991: 165). Not only; whenever we speak of 'space' in myth, it is the *world* that we speak of. This is exactly what Hübner means when he notes that mythical space is not some universal realm within which objects are found but, rather, that space and its contents form one inseparable whole. In Hübner's eyes, this understanding represents the *first* key difference between scientific (or cartographic) conceptions of space and mythical ones. A *second* difference, according to Hübner, lies in the fact that within these latter, space is not simply composed of a continuum of a multiplicity of points but is, rather, made up of discrete elements, the so-called *teméne*, which through their alignment constitute the spatial dimension. The *third* difference lies in the fact that mythical space is never homogeneous, since places have both a relative as well as an absolute position. The *fourth* difference is that mythical space is not isotropic, since subject to differentiation dependent on the *direction* of events. The *fifth* difference pertains to the distinction between sacred and profane spaces. The former are, most often, parts of the latter, although this is not true of all sacred spaces. Some, in fact, are 'prime' spaces (in the mathematical sense of the word) and thus irreducible in space; others, while also such 'prime' spaces, can be found repeatedly within secular spaces. In topological terms, profane spaces are bound by just such a dual logic and, therefore, distinguishable only by the fact that sacred spaces, characterized as they are by discontinuity and a lack of homogeneity and isotropy, are either unreachable or appear under a multiplicity of forms, although they are one and the same place. The *sixth* difference is that in myth, there is no singular space within which everything has its place, within which all things are part of an overarching order. Myth is dominated, rather, by the juxtaposition of singular elements (that which quantum theory refers to as 'entanglement'), elements which are at times sacred and at times profane (as are the shifting states theorized by quantum theory). The *seventh* and final difference is that profane spaces are metrically defined, as each of their component objects is endowed with a three-dimensional metric. This rule, however, does not hold for sacred spaces (Hübner 1990: 182–3).

Returning to Castells' (1996: 423–8) analysis, we can note that one of the necessary albeit not sufficient conditions for distinguishing the space of places from the space of flows is precisely the presence of a metric and relations of contiguity, evident within the former while not in the latter. In other words, the relation between the space of flows and the space of places is much like the one between sacred and profane places specified above. In fact, the net which both differentiates between as well as connects the two spaces functions precisely because it, too, is governed by two of the characteristics of that which Ernst Cassirer (1964: 55–7, 73–5) has termed the "mythical consciousness of the object": (i) the absence of any distinction between a whole and its parts conceiving, rather, each part as the totality (not merely in symbolic or conceptual terms but in objective, tangible ones), and (ii) the absence of any distinction between image and subject, in the sense that within mythical consciousness, the image does not merely represent the object but, rather, *is* the object. If we agree with the above, we can argue that both modernity (within which the cartographic image of the world *is* the world) as well as postmodernity possess a mythical nature, albeit within this latter "issues of space have been supplanted by a kind of immersion of the Self in the mediascapes of teleculture which must generate a communicative practice whose boundaries are mapped in virtual, transitory networks, whose hold on matter is ephemeral, whose position in space is tenuous and whose agency is measured in act of implication rather than mere coincidences of location" (Druckrey 1996: 20).

But we should return here, again, to the earliest stirrings of Western culture and the first realization that any such coincidences could never be perfect. I'd like to turn to Hübner's (1990: 183) work again in order to explain to any readers not well acquainted with the principles of topology just what I intend. Hübner presents the example of the stereographic projection of a sphere on a plane, where for every point P_1 on the sphere there exists a correspondent point P_1 mapped onto the plane. What he notes is that this projection necessitates a dismemberment of the entire set of points which make up the volume of the sphere. Yet a certain point P (which corresponds roughly to the North Pole) cannot possibly correspond to *any point* on the plane, for its projection lies at infinity. The same is true, in fact, of point O which is situated at the opposite, inferior pole, as it constitutes the only point of contact between the sphere and the plane. Here, too, projection is impossible. What Hübner does not stress, however, is that the position of these two points – and thus their untranslatability – renders them sacred: the first is none other than Olympus, the second, Tartarus. *The sacred is therefore that which evades the reduction of the globe to a map.*

Heisenberg (1960: 53–4) once wrote that any attempt to understand life through the principles of classical physics was just as misguided as the vain hopes of a wanderer who believes (s)he could solve the enigmas of the world by traveling to its furthest confines. Few other passages provide such a concise and forceful pronouncement of the end of modernity – granted that we are able to recognize the above wanderer as none other than the Renaissance *Homo Viator*, Descartes's 'strange soldier' – and that within the principles of classical physics we read the most forceful and coherent expression of the cartographic ethos. It is this ethos that we must abandon in order to undertake our postmodern journey. With the certainty that Einstein was right on one point in his attack on the Copenhagen School: "God does not throw dice" – but only because this would imply the existence of a table, of a map. God, however, created a sphere, a globe.

References

Auyang, S. Y. (1995). *How is Quantum Field Theory Possible?* Oxford: Oxford University Press.

Barglow, R. (1994). *The Crisis of the Self in the Age of Information: Computers, Dolphins, and Dreams.* London: Routledge.

Baudrillard, J. (1981). *Simulacres et simulation.* Paris: Galilée.

Birkoff, G. and J. von Neumann (1936). The Logic of Quantum Mechanics. *Annals of Mathematics* 37: 823–43.

Blumenberg, H. (1991). *Elaborazione del mito.* Bologna: Il Mulino.

Bohr, N. (1965). *I quanti e la vita.* Turin: Bollati Boringhieri.

Briggs, R. A. (1987). Quantum Theory and Geography. *Earth Surfaces Processes and Landforms* 12: 571–3.

Burckhardt, J. (1980). *La civiltà del Rinascimento in Italia.* Florence: Sansoni.

Cassirer, E. (1964). *Filosofia delle forme simboliche. Il pensiero mitico,* vol. 2. Florence: La Nuova Italia.

Castells, M. (1989). *The Informational City: Information, Technology, Economic Restructuring and the Urban-Regional Process.* Oxford: Blackwell.

Castells, M. (1996). *The Information Age: Economy, Society and Culture: The Rise of Network Society,* vol. 1. Oxford: Blackwell.

Chevalley, C. (1984). Une nouvelle science. In *Le monde quantique,* ed. S. Deligeorges. Paris: Seuil, 33–50.

Claval, P. (1971). *L'evoluzione storica della geografia umana.* Milan: Franco Angeli.

Collier, P., C. Dewdney, R. Inkpen, H. Mason, and D. Petley (1999). On the Non-necessity of Quantum Mechanics in Geomorphology. *Transactions of the Institute of British Geographers* 24: 227–30.

Conley, T. (1997) *The Self-Made Map: Cartographic Writing in Early Modern France*. Minneapolis: University of Minnesota Press.

Dalla Chiara, M. L. and R. Giuntini (1997). La logica quantistica. In *Filosofia della fisica*, ed. G. Boniolo. Milan: Bruno Mondadori, 609–45.

Druckrey, T. (1996). Introduction. In *Electronic Culture: Technology and Visual Representation*, ed. T. Druckrey. New York: Aperture, 12–25.

Farinelli, F. (1976). La cartografia della campagna nel Novecento. In *Storia d'Italia*, VI, *Atlante*, ed. L. Gambi. Turin: Einaudi, 626–54.

Farinelli, F. (1981a). Dallo spazio bianco allo spazio astratto: la logica cartografica. In *Paesaggio: immagine e realtà*, ed. T. Maldonado. Milan: Electa, 199–207.

Farinelli, F. (1981b). Il villaggio indiano o della geografia delle sedi. Una critica. In *Il villaggio indiano. Scienza, ideologia e geografia delle sedi*, ed. F. Farinelli. Milan: Angeli, 9–50.

Farinelli, F. (1983). Introduzione ad una teoria dello spazio geografico marginale. In *L'Italia emergente. Indagine geo-demografica sullo sviluppo periferico*, eds. C. Cencini et al. Milan: Franco Angeli, 17–32.

Farinelli, F. (1989a). *Pour une théorie générale de la géographie*. Genève: Département de Géographie de l'Université.

Farinelli, F. (1989b). Certezza del rappresentare. *Urbanistica* 97: 7–16.

Farinelli, F. (1990). Dancing. In *La transformation de l'environnement quotidien: représentations et pratiques*, ed. W. Leimgruber. Fribourg: Institut de Géographie de l'Universitè, 1–14.

Farinelli, F. (1992). *I segni del mondo. Immagine cartografica e discorso geografico in età moderna*. Florence: La Nuova Italia.

Farinelli, F. (1995). L'arte della geografia. *Geotema* 1: 139–55.

Farinelli, F. (1997a). Von der Natur der Moderne: eine Kritik der kartographischen Vernunft. In *Räumliches Denken*, ed. D. Reichert. Zürich: Hochschulverlag AG an der ETH.

Farinelli, F. (1997b). L'immagine dell'Italia. In *Geografia politica delle regioni italiane*, ed. P. Coppola. Turin: Einaudi.

Farinelli, F. (1998a). Did Anaximander ever Say (or Write) any Words? The Nature of Cartographical Reason. *Ethics, Place and Environment* 2: 135–44.

Farinelli, F. (1998b). Il pappagallo degli Atures. In *Quadri della Natura*, by A. von Humboldt. Florence: La Nuova Italia, vii–xxvi.

Farinelli, F. (1999a). Lo sguardo di Guatarrale, il silenzio di Kant, gli occhi di Humboldt. In *La costruzione del paesaggio siciliano: geografi e scrittori a confronto*, ed. G. Cusimano. Palermo: Facoltà di Lettere e Filosofia dell'Università, 147–54.

Farinelli, F. (1999b). La globalizzazione. *I viaggi di Erodoto* 40: 18–27.

Florenskij, P. (1984). *La prospettiva rovesciata e altri scritti*. Rome: Casa del Libro.

Ghirardi, G. C. (1997). I fondamenti concettuali e le implicazioni epistemologiche della meccanica quantistica. In *Filosofia della fisica*, ed. G. Boniolo. Milan: Bruno Mondadori, 337–608.

Gusdorf, G. (1984). *Mythe et métaphysique. Introduction à la philosophie*. Paris: Flammarion.

Haggett, P. (1965). *Locational Analysis in Human Geography*. London: Arnold.

Harrison, S. and P. Dunham (1998). Decoherence, Quantum Theory and their Implications for the Philosophy of Geomorphology. *Transactions of the Institute of British Geographers* 23: 501–14.

Harrison S. and P. Dunham (1999). Practical Inadequacy or Inadequate Practice? Quantum Theory, 'Reality' and the Logical Limits to Realism. *Transactions of the Institute of British Geographers* 24: 236–42.

Hartshorne, R. (1961), *The Nature of Geography*. Lancaster, Penn.: The Association of American Geographers.

Heisenberg, W. (1960). *Mutamenti nelle basi della scienza*. Turin: Boringhieri.

Heisenberg, W. (1961). *Fisica e filosofia. La rivoluzione nella scienza moderna*. Milan: Il Saggiatore.

Hübner, K. (1990). *La verità del mito*. Milan: Feltrinelli.

Humboldt, A. (1845). *Kosmos. Entwurf einer physischen Weltbeschreibung*, vol. 1. Stuttgart u. Tübingen: Cotta.

Husserl, E. (1961). *La crisi delle scienze europee e la fenomenologia trascendentale*. Milan: Il Saggiatore.

Kelly, K. (1995). *Out of Control: The Rise of Neo-biological Civilization*. Menlo Park, Calif.: Addison-Wesley.

Lefebvre, H. (1978). *La produzione dello spazio*, vol. 1. Milan: Moizzi.

Lévi-Strauss, C. (1930). Préface. In *Sociologie et Anthropologie*, V–LVI, ed. M. Mauss. Paris: Presses Universitaires Françaises.

Lévi-Strauss, C. (1962). *La pensée sauvage*. Paris: Plon.

Lindley, D. (1997). *La luna di Einstein*. Milan: Longanesi & C.

Maffesoli, M. (1985). *L'ombre de Dionysos. Contribution à une sociologie de l'orgie*. Paris: Klincksieck.

Nebel, G. (1961). *Pindar und die Delphik*. Stuttgart: Klett.

Olsson, G. (1991). *Linee senza ombre. La tragedia della pianificazione*. Roma-Napoli: Theoria.

Olsson, G. (1998). Towards a Critique of Cartographical Reason. *Ethics, Place and Environment* 2: 145–55.

Pecora, A. (1970). La 'corte' padana. In *La casa rurale in Italia*, eds. G. Barbieri and L. Gambi. Florence: Olschki, 203–21.

Popescu, S. and D. Rohrlich (1997). Action and Passion at a Distance: An Essay in Honour of Professor Abner Shimony. In *Potentiality, Entanglement and Passion-at-a-Distance*, vol. 2, eds. R. S. Cohen, M. Horne, and J. Stachel. Dordrecht-Boston-London: Kluwer, 197–206.

Ptolemy, *Geography*.

Reichenbach, H. (1957). *The Philosophy of Space & Time*. New York: Dover Publications.

Ritter, C. (1852). *Einleitung zur allgemeinen vergleichenden Geographie und Abhandlungen zur Begründung einer mehr wissenschaftlichen Behandlung der Erdkunde*. Berlin: Reimer.

Schmitt, C. (1981). *Land und Meer. Eine weltgeschichtliche Betrachtung*. Köln-Lövenich: Maschke-Hohenheim.

Shimony, A. (1993). *Search for a Naturalistic World View: Natural Science and Metaphysics*, vol. 2. Cambridge: Cambridge University Press.

Spedding, N. (1999). Schrödinger's Cat Among the Pigeons? *Transactions of the Institute of British Geographers* 24: 231–5.

Troeltsch, E. (1989). *Lo storicismo e i suoi problemi*. Naples: Guida.

van der Krogt, P. (1993). *Globi Neerlandici: The Production of Globes in Low Countries*. Utrecht: Hes Publishers.

van der Leeuw, P. (1945). *L'homme primitif et la religion*. Paris: Presses Universitaires Françaises.

Volk, T. (1995). *Metapatterns: Across Space, Time, and Mind*. New York: Columbia University Press.

Wittkower, R. (1985). *La scultura*. Turin: Einaudi.

Zurek, W. H. (1993). Preferred States, Predictability, Classicality and the Environmental-induced Decoherence. *Progress of Theoretical Physics* 89: 281–312.

Zurek, W. H. (1994). Preferred Sets of States, Predictability, Classicality and Environment-induced Decoherence. In *Physical Origins of Time Asymmetry*, eds. J. J. Halliwell et al. Cambridge: Cambridge University Press, 197–206.

12

Washed in a Washing Machine™

Gunnar Olsson

Telling the truth is easy, being believed a simple trick. Why, then, is it so difficult – perhaps impossible – to tell the truth and be believed at the same time?

In the 1920s Niels Bohr and his fellow physicists were faced with essentially the same problem of how to bring the classical notions of causality and the divisibility of space and time into harmony with the notions of probability, discontinuity, and wholeness characteristic of atomic processes. The difficult question is this: how can I remain sane and still perform the multiple and contradictory roles of actor and spectator, exhibitionist and voyeur?

The short answer is that our thoughts-and-actions are governed by a form of cartographical reason, a mode of understanding which simultaneously tells me where I am and where I should go; to be sane is to know one's way, to be crazy is to be lost. Nice and clear, until I discover that the believable map and the meaningful world are not one and the same; while the former is a flat surface hung on the hook of the magnetic north pole, the latter is a folded globe without fix-points. The (post)modern reality of time–space compression is not what it seems to be, not a bouncing ball of rubber but a sturdy Henry Moore construction of big holes and solid connections.

To these confusions should be added that in every *already* there is a *not yet*, an inbuilt quest of another place and time, a movement from the now-here of the Lacanian real to the no-where of the utopian. The world does not sit still and major parts of human life are literally untouchable. Trans-lation is nevertheless the name of the game, for to translate is by definition to make new boots out of old ones, to convey an idea from one art form to another, to play with the expression of *Ceci n'est pas une pipe*.

Ceci n'est pas une pipe.

Figure 12.1 René Magritte, *La trahison des images (Ceci n'est pas une pipe)*, 1928/9, by permission of the Magritte estate and the Los Angeles County Museum of Art. Purchased with funds provided by the Mr. and Mrs. William Preston Harrison Collection. © ADAGP, Paris and DACS, London, 2001. Photograph © 2000 Museum Associates/LACMA

Through his famous painting of this title, René Magritte provided a preview of what we nowadays define as the self-referential problem of (post)modern geographical praxis. For just as the Belgian surrealist through his geometrical experiments struggled with *The Treason of Pictures*, so a handful of geographers have been approaching the same boundary-zone between semiotics and politics. It is in fact when Magritte's statement is read as a map and not as a painting that it best illustrates how we find our way in the unknown, in the human realm of love and trust as well as in the physical world of mountains and rivers. Yet another show of the cultural relations between thing and meaning, body and mind. As solid rocks turn to quicksand, geometry loses its footholds, the compass its direction.

An unexpected virtue of Magritte's canvas is that it indicates exactly where the present collection of papers can be located: in the grayish space which fills the void between the picture of the pipe and the story of the words. To be more exact, the surveyor's fix-points are lodged in the concept of the *Ceci*, in the grammar of the 'this,' in the geographical intertwining of the two prepositions 'of' and 'in.' As Samuel

Figure 12.2 Albrecht Dürer, *The Designer of the Lying Woman*, print from Dürer's textbook *Unterweysung der Messung mit dem Zickel und Richtscheyt*, 1538

Beckett once remarked about James Joyce, the challenge is not to write *about* something; it is to produce a text which at the same time *is* that something. Magritte's *Ceci* in another context.

*   *   *

Speeding up, social cloning. Wheels spinning, intellectual mutation. The history of (post)modernism is always in the making, its final destination inherently unknowable, its searching beginnings clearly visible in the records of the Italian Renaissance; deeply embedded in Filippo Brunelleschi's revolutionary invention of the self are the shifting relations between the vanishing point of the picture, on the one hand, and the grounded viewpoint of the spectator, on the other. As masterly demonstrated by Hubert Damisch, there runs a thick (albeit crooked) line from the Ideal City of the fifteenth century Urbino panel to the real city of twentieth-century Los Angeles, from the single focus of Piero della Francesca to the reflecting mirrors of Jacques Lacan. The *not yet* of Geography is foreshadowed by the *already* of the visual arts.

One century after Brunelleschi, Albrecht Dürer experimented with the same principles of representation in his famous wood-cut *The Designer of the Lying Woman*. Caught in the orthogonal coordinate net of the vertical window is the two-dimensional projection of the three-dimensional object placed on the other side of the glass, a paradigmatic illustration of cartographical reason. Thus, as the gridded image on the glass is transferred first to the artist's pen and then onto his paper, the outcome is a classical map, seemingly of the voluptuous nude, in reality of the self-referential practice of mapping itself. The bride stripped bare by the Christ-like Dürer four centuries before Marcel Duchamp performed his own version of the same act. Have another apple, my dear! And behold, your eyes will be opened, the fig-leaf designers forthcoming.

It is in these performances at the outer limits of projective mapping that the copies and originals of (post)modernism come naked before us, issues of re-presentation revealing themselves as issues of re-creation. Driving the striptease is an irresistible urge to reach beyond the surface level of the taken-for-granted, to suck yet another bite of the unconscious into the realm of the conscious. Nothing obscene about that, merely an alternative way of operationalizing the well-established epistemology of extremity, an unexpected parallel to the mathematician's procedure of letting his functions approach zero and infinity.

When the tongue slips, it bares the teeth of its owner, no dentist strong enough to pull them out. And yet, both zero and infinity are recent inventions, foreign orphans adopted by the Renaissance.

*

No one has explored the Land of the Taken-for-Granted with greater success than Marcel Duchamp. If anyone, this leading practitioner of delays learned that in every already there is a not yet. If anyone, this timely person knew that ideas are mental constructs without form and void. And for exactly these reasons he devoted his life to studies of the geometry of the invisible. To be a genius is to know the impossible.

The turning-point of Duchamp's career occurred in 1912, when the 25-year-old reached the conclusion that if painting was to be taken any further, someone else would have to do it. Hence he abandoned painting, not because he had lost interest in the arts or in the limits of representation, but because he was determined to go beyond the confines of the one-point perspective. With characteristic honesty he told himself, "Marcel, no more painting, go get a job." Which he did, albeit only for a year and a half.

When reaching this decision, the young man had just returned to Paris from a formative period in Munich, where he had continued his earlier struggle with the transitions of cubism, especially with the problem of how to picture the passage from virgin to bride. In addition, he bought a copy of Wassily Kandinsky's new book *Über das Geistige in der Kunst*. His real obsession was to go beyond his own *Nu descendant un escalier*, the painting which for him represented the definitive break with realism. As he put it in a later interview (Cabanne 1977: 35), "when you want to describe an airplane in flight, you don't paint a still life."

The story is well known, not the least through the informative catalog from the 1993 exhibition at Palazzo Grassi. Thus, when Duchamp took his painting to the Salon des Indépendants in Paris, he sincerely expected

Figure 12.3 Marcel Duchamp, *Nu descendant un escalier, No. 2*, 1912, reproduced by kind permission of the Philadelphia Museum of Art, Louise and Walter Arensberg Collection; © Succession Marcel Duchamp/ADAGP, Paris and DACS, London 2001

it to be welcome in the cubist room. But on Monday March 18, 1912, he receives a surprise visit from his two brothers, the painter Jacques Villon and the sculptor Raymond Duchamp-Villon, dressed as for a funeral. Their attire was appropriate, for they were sent by the hanging committee to inform their younger relative that his work was considered unfitting. One opinion was that the submitted canvas had "too much of a literary title," another that "a nude never descends the stairs – a nude reclines."

"The cubists think it is a little off beam," said the brothers. "Couldn't you at least change the title?" But as the title was written in capital letters at the bottom of the canvas, it is an integral part of the painting itself. Hence it could not be removed. Magritte's *Ceci* 14 years before the pipe.

Marcel said nothing. But immediately the messengers had left, he took a taxi to the Quai d'Orsay, collected his picture, and brought it home.

The betrayal was never forgotten, the world irrevocably changed in the process – an Amazon butterfly fluttering its wings in a Paris building with water and gas on every floor. Just as the cubist nude did not fit into Dürer's orthogonal coordinate net, so the ambitious Marcel did not conform to the establishment definition of what it is to be a painter. The wounds never healed although he was later able to recall (Cabanne 1997: 60) that the affair "helped me to free myself completely from the past, in the personal sense of the word. I said to myself: right, since that's how things are, I'll never join a group again, I have to rely on myself, I have to go it alone." As a consequence, he came to spend the next 11 years working on the *Large Glass*, his own definitively unfinished study of Dürer's fresh wi(n)dow, his personal deconstruction of the Renaissance perspective, the scathing response to his insensitive critics. The bachelor was forced to grind his own chocolate, albeit in the company of Francis Picabia and under the influence of Raymond Rousel.

Eleven months after the *Nu descendant un escalier* had been rejected in Paris (and after brief showings in Barcelona and Paris), it was exhibited anew, this time at the Armory Show on Lexington Avenue in New York. The reception of the work, now under the domesticated title *Nude Descending a Staircase*, was beyond belief. In his physical absence, Duchamp was instantly turned into a celebrity, the only Frenchman as well known to the American public as Napoleon and Sarah Bernhardt. (Post)modernism in advance of itself.

*

At the time of the Armory Show, Duchamp had already given up painting, allegedly for good. On the newly plastered wall of his Paris apartment, 23 Rue Saint-Hippolyte, he was instead busy plotting the full-scale composition of what years later was to be the *Large Glass*; the focal point of the Bachelor Machine was exactly defined as early as January 1913. But it was not until 1915, after he had moved to New York, that he created the concept of a readymade, by definition an article of everyday use so trivial that it normally goes unnoticed. Towards the end of his life (Duchamp 1968: 47), he explained how "a readymade is a work of art without an artist to make it . . . A tube of paint that an artist uses is not made by the artist, it is made by the manufacturer that makes paints. So the painter really is making a readymade when he paints with a manufactured object that is called paint." Since the word 'art' means 'making' and 'making' means 'choosing,' the choice of a readymade is analogous to the choice of a tube of paint. In exactly the same sense, every word of the present text is a readymade, an anonymous entity twisted, turned, loved, and hated by some choosy author(itarian). We never begin from scratch.

In that molding process, the *not yet* is brought about through the same two-stage procedure by which God originally created the world: first an object is removed from its ordinary context and placed in another, then its name is changed to fit better into the new surroundings. The pivotal moment came in the spring of 1917, when the resident alien otherwise known as Henri-Robert-Marcel Duchamp, originally born on July 28, 1887, near Blainville, France, went into a New York hardware store, bought the infamous porcelain urinal, carried it to the exhibition hall of the Society of Independent Artists, turned it 90 degrees, signed it "R. Mutt" and baptized it *Fountain*. The scandal was immediate, the thing itself first hidden behind a screen, then rejected and eventually lost. Its non-existence notwithstanding, the piece was projected into eternity through a photograph taken by the trend-setting photographer Alfred Stieglitz, well aided by some spicy anecdotes strategically planted by Duchamp himself. Sacrificed was the material object, resurrected was the untouchable thought.

The story of the *Fountain* closely resembles the story of the *Nude Descending a Staircase*. Once again it was made clear that art is not a thing but a statement, that making is an act of choosing; an excellent example of the semiotic intertwining of Signifier and signified, the One turning into the other, the Other into the one. The magic wand of ontological transformations is forged through language, the performance itself a show of rhetorical tropes. Let there be! And there is. Four cases to illustrate the point (de Duve 1996: 398):

1 the hat rack exhibited in 1917 and quickly lost, its fate effectively written into the title itself: *Hat Rack*. A tautology, by definition always true but never informative. Bertrand Russell's theory of proper names and definite descriptions rejected by a Frenchman who had never heard of it;
2 the coat rack which Duchamp initially had intended for the vertical wall of his New York studio but eventually nailed to its horizontal floor instead; as he later recalled (d'Harnancourt and McShine 1973: 283), "a real coat hanger that I wanted sometime to put on the wall and hang my things on but I never came to that – so it was on the floor and I would kick it every minute, every time I went out – I got crazy about it and I said the Hell with it, if it wants to stay there

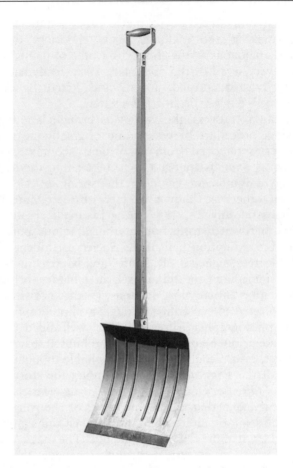

Figure 12.4 Marcel Duchamp, *In Advance of the Broken Arm*, 1916, Replica 1964, Yale University Art Gallery. © Succession Marcel Duchamp/ADAGP, Paris and DACS, London 2001

and bore me, I'll nail it down . . . It was not bought to be a Ready-made – it was a natural thing . . . it was nailed where it was and then the idea came . . ." In the same manner as the urinal had been renamed *Fountain*, so the ordinary coat rack was now baptized *Trébuchet*, a French chess term which means *Trap* or *Stumbler*. A textbook application of metaphor, the rhetorical trope by which physical likeness is translated into the nearness of metonymical associations;

3   the bicycle wheel mounted by its fork on a painted wooden stool (original [Rue Saint Hippolyte, Paris, 1913] lost; 2nd version [33 West 67 Street, New York, 1914] lost; 3rd version made for 'Chal-

lenge and Defy,' Sidney Janis Gallery, New York, 1951; 4th version Ulf Linde, Galleri Burén, Stockholm, 1961; 5th version Richard Hamilton, London, 1964; 6th version, edition of 8 signed and numbered replicas, Galleria Schwartz, Milan, 1964). Even though its title, *Roue de bicyclette*, (literally but flatly translated as *Bicycle Wheel*) sounds like a stuttering tautology, it is in effect a meaning-filled synecdoche, the name of a part readily extended to the whole. Since the *Bicycle Wheel* is not a pure but an assisted readymade, the chosen trope is highly appropriate;

4   the beautiful snow shovel exhibited in 1915, subsequently returned to its daily chores and gone for ever. For who, except a crazy avant-garde artist, could imagine that the snow shovel (wood and galvanized iron, 121.3 cm high) was not a snow shovel but an allegory? *In Advance of the Broken Arm*. But why 'arm' rather than 'back' or 'heart'?

*

Duchamp's linguistic and rhetorical skills, his punnings and humorous word-plays, were incomparable; he was certainly not a person to fall into the political trap of confusing the material *urbs* with the cultural *civitas*. I might well be the only person tempted to call him a (post)modernist, but I have long been struck by the many similarities between Duchamp's paradoxical juxtapositions, on the one hand, and the textual inventions of writers like Ihab Hassan, Octavio Paz, William Gass, Gabriel García Márquez, Julio Cortázar, and Thomas Pynchon, on the other. How do I weave the three threads of rendered, renderer, and rendering into the same seamless web? How do I react when I realize that the water is more important to the fish than the fish to the water?

But there is a hole in the bucket. And that leakage explains why the bright hopes which initially were associated with the term 'postmodern geographical praxis' have all but disappeared. Quite understandable, though, and for two closely related reasons: firstly, the principle of self-reference (*the* principle that fuels the engine of [post]modernism) violates some of the most sacred beliefs of scientific methodology; secondly, the (post)modern cultivation of the fragment runs counter to almost everything which is currently considered politically correct. My own conclusion is that serious (post)modern praxis is far too radical for the majority of self-appointed radicals. Why else would there have been so few attempts to experiment with the taken-for-grantedness of the printed text, even fewer to enter the no-man's land between body and meaning? Very little of Kierkegaardian pseudonyms, very much of dyslexic writers, very little of swarming pathways, very much of one-way streets. As the politics of the Other is degenerating into the politics of the Same, difference is banned, identity fostered.

Perusing the academic Geography presently in vogue, I am often reminded of the Jesuit missionaries who tried to force the syntax of Japanese into the grammar of Latin. Allan Pred and Marcus Doel are rare exceptions. Pricey prizes placed on their heads.

*   *   *

I am personally convinced that the young geographer interested in the power of the gaze and the politics of the Other has more to learn from recent art history than from the disciplining masters of Vancouver and Los Angeles. The decisive difference is that while the latter imagine that they are looking at the world, the former know that the world is looking at them. At issue is the issue of engagement.

And thus it was that Paul Cézanne a century ago could prepare the way for the coming revolutions in painting. Since this antisocial son of a self-made banker never took anything for granted, he was gradually able to acknowledge that he no longer painted God-given landscapes but man-made pictures, not houses and mountains but triangles and rectangles, not content but form. Transcended in the process was the Renaissance perspective, that single point of vision which for half a millennium had stabilized what people saw. In Cézanne's own words, he sought to render perspective only through color. His later canvases became almost abstract.

Rather than relying on classical perspective and modeling with line and shadow, Cézanne used planes of color, not the least in his repeated attempts to capture the landscape of Mont Sainte-Victoire. What he thereby learned was not only that colors get darker the farther they are from the viewer, they change in shade as well – when orange recedes it turns to red, when yellow moves away it fades into green. As a further development he began to slant the patches of color in different directions, composing a series of paintings of three-dimensional solidity and depth in the process; the seemingly compact objects are produced through strokes which in themselves are open forms. The effects that Brunelleschi and Dürer attained by making far-away objects smaller (the social gravity model of $Iij = f(Dij)$ comes readily to mind), Cézanne achieved by changing the color. His secret was to take the word 'painting' literally, to see that when color is at its richest, form is at its bodily limit. There is a new kind of empiricism, naive and deeply sincere, eventually hermetic.

Nothing new under the sun, though. Always an *already* in the *not yet*. And since Venice is Venice, it is impossible not to recall Jacopo Tintoretto, the genius whose name was one with his ancestry: the dyer's son. For it was he, perhaps earlier than anyone else, who attained a sense of depth by working directly with the light of color, allowing the central focus of the vanishing point to split into many. In these twisting and unstable figures of the mannerist, there is a foreshadowing not only of the folds and ovals of the Baroque but also of the thousand plateaux and lines of flight associated with the multicentered universe of Gilles Deleuze and Félix Guattari. Even though the typical (post)modernist is a schizophrenic, he is not crazy. Even though the trendy jet-setter is a nomad, he is a monad as well.

Sensing the parallels between Tintoretto, Leibniz, and the (post)modern, I am reminded of El Greco too, especially of *The Burial of Count Orgaz*. In my mind, an almost perfect depiction of the Saussurean sign of

$$\frac{\underline{s}}{S}$$

In the lower, earthly, portion of this painting lies the concrete body of the imaginary Signifier (S), in the upper, heavenly, portion flows the abstract meaning of the symbolic signified (s). In-between is the fraction line of the Bar, that magic wand which El Greco filled with living people from the real city of Toledo, including a fine self-portrait of himself and a touching picture of his son. The signature is on a handkerchief.

*

Every artist knows that painting is a study of color. Taking painting seriously is therefore to take the paint seriously, to treat the manufactured paint as a readymade. By extension, the critical social scientist has no choice but to take the geography of grammar seriously, to approach this word *Ceci* as the readymade it actually is. But just as the amateur painter rarely sees the things as they really are, so the majority of academic writers rarely read the words as they really mean. In contrast, the professional inmates of the prison-house of language know that grammar is a map of structural repression, hence of individual escape routes as well; in the sensitive hands of these artists, nouns become files for cutting the bars, verbs turn to ladders for climbing the walls, prepositions get transformed into guardians as prone to killing as to looking the other way.

For these reasons of grammar it is rewarding to make a brief detour into the peculiarities of the Finnish language. To most people (including myself) a strange and bewildering idiom, not the least because it fails to classify nouns according to gender. It follows that the Finnish speaker has no direct way of distinguishing between a 'he' and a 'she.' For the gossipmonger a heaven of creative ambiguities, for the feminist a hell

Figure 12.5   El Greco, *The Burial of the Count of Orgaz*, 1586, San Tomé, Toledo

of confusing categories. When listening to a native Finn speaking Swedish or English, the unaware is easily lost. Interesting in deed.

And yet. Given the present context, the most remarkable aspect of the Finnish language is not the absence of gender but the presence of different cases – 15 in total – a powerful means for marking directly on the noun where the subject and object are positioned in relation to each other. Finnish authors are consequently well equipped to do with their words what Cézanne did with his palette and Brunelleschi by moving his easel. The classical example is Veijo Meri's nontranslatable novel *Peiliin piirretty nainen*, in Swedish interpreted as *Kvinna i spegeln*, literally *Woman in the Mirror*. Through an intricate play of word-endings, the author is here able to specify directly, rather than circumstantially, whether the woman in question is *at* the mirror, *on* the mirror, *in* the mirror, *in front of*, or *out of* the mirror. Cubism by another route, the picturer of a pipe transformed into a piper of words.

In most Indo-European languages, prepositions perform essentially the same functions as the cases do in the Uralic language family; in Western thought-and-action no part of speech is more crucial than the preposition. Indeed there is a choreography of Power beautifully reflected in the movements of the word 'power' itself. For such is its nature that Power never stays put, always enters the Saussurean Bar disguised and incognito. On the surface, a concrete noun of whips and guns, fists and shouts, deeper down a noun of abstract relations; every 5-year-old knows that although sticks and stones may break her bones, it's the words that really hurt her. For Power is not merely a thing, but a special type of action as well. And as the drama develops, the word 'power' turns into a verb; transitive, intransitive, and – most notably – a range of modal auxiliaries. There is a 'will' to Power, a tempting relation between the possible stands of 'may,' 'can,' 'ought,' 'shall,' 'must,' etc.; Friedrich Nietzsche is best described as a perspectivist.

But just as a particular thing may be beautiful or ugly, so a particular act may be better or worse; Kant's *Critique of Judgment* is more basic than his previous critiques of pure and practical reason. Speaking directly to that point, a range of adjectives and adverbs briefly enter the stage, immediately followed by a troop of dancing pronouns and a choir of prepositions and conjunctions. In the ensuing turmoil, the most common expressions are nonsensical interjections. "Oh, my dear. Really? Very well! Of course. A fact, absolutely. Certainly. Sure. Oh, yes. Him as well as another." The curling of thought. Sand on the ice, please!

And as the slowed-down play approaches its climax, prepositions become the leading actors. Power stands naked before us, undressed on the balcony, its Dyonisian mask removed. Hierarchies *do* matter, for even though the ontological magician performs his tricks in the tabernacle of a traveling theater, the actor's position on the stage is carefully circumscribed; there are obvious and crucial differences between being *at* a limit, *on* a limit, *in* a limit. Pre-positions are in effect thoughts-and-actions in advance of themselves, the taken-for-granted *in nuce*, a snow shovel in disguise. It says a lot about the spatial structure of the taken-for-granted that prepositions are the most difficult words to master in a foreign language.

A century past, not wasted. For now it can be said explicitly: (post)modern praxis has never again reached the same heights as in the hazardous poetry of Stéphane Mallarmé (*Un coup de dés jamais n'abolira le hasard*, 231–3):

N'AURA EU LIEU QUE LE LIEU

EXCEPTÉ

PEUT-ÊTRE

UNE CONSTELLATION

Toute pensée émet un Coup de Dés

*

It is difficult to conceive of painting without paint, even harder to imagine painting without a surface onto which the paint is deposited. The *already* of that problem goes far back, at least to the time of Plato's *Republic*, especially to the story of the cave. Not, however, to its comments about the shining Sun and the chained prisoners, but to its remarkable ignorance of the wall. The silence is stunning, for to us it seems self-evident that without the resisting wall there would be no shadows at all, hence nothing to see, hence nothing to share either. Understanding the structure of the projection plane is a challenge of the utmost order.

To be metaphorically precise: not all paint sticks to all surfaces, for just as some mixtures run off without leaving a trace, others crack as they harden. To prepare the canvas for the unknown, the artist tries to kill it with layers of gesso, a practice highly reminiscent of the socialization processes through which you and I are made so obedient and so predictable. The parallel is obvious, for just as some readers will immediately grasp the present text, others will neither understand nor remember. In through one ear, out through the other. No traces left in between. Not with **you**, however! Because **you** are different, **you** are still with me, the proof beyond doubt: it is **you** – and **you** alone – who is reading this sentence. Not in an imaginary and free-floating No-Where, but in the real presence of the Now-Here. Just as the painter's paintbrush leaves its marks on the stretched and prepared canvas, so my delayed wordbrush touches the wrinkled surface of your socialized unconscious. In both instances, the signified springs from the meeting of a fleeting Signifier and a capturing screen.

When Kazimir Malevich chose to paint a square rather than a pipe, it was this convergence of image and taken-for-granted that he wished to understand. As noted elsewhere (Olsson 1994), the *Black Square* of

Figure 12.6   Kazimir Malevich, *Black Square*, 1929, reproduced by kind permission of The State Tretyakov Gallery, Moscow

1914 was in fact an icon. But the truth of an icon is not what it appears to be, for its secrets are not in the light-waves that meet the eye, but in the meanings that stir the mind. With its antecedents in Egyptian mummy-portraiture, an icon is not a picture of the holy, it is holy in itself. It follows that the *Black Square* is effectively a graven image, a godhead made of four namable points, four silent lines, and one unspeakable plane. While three of the corners are conventionally baptized 'Father,' 'Son,' and 'Holy Spirit,' the fourth takes its name from Walt Whitman's 'Rebel,' 'Satan,' or 'Comrade of Criminals.' Even more revealing, though, is Malevich's canvas from 1918, *White Square on White*, now in the possession of the Museum of Modern Art in New York.

What makes this latter work so crucial is that the icon of the tilted white square is set against a background which itself is white. It follows that the painting at the same time is a picture of the nonpicturable and a study of the screen onto which the forbidden image is cast. For the cartographer of Power a beautiful instance of an invisible map of the invisible: the taboo-ridden taken-for-granted captured in the language of the same; a picture of a dematerialized object; a story of self-reference. Screaming silence, difference deferred. And, as if to illuminate its own point, the *White Square on White* is virtually impossible to reproduce. For what speaks in Malevich's canvas is not the color of the paint but the texture of the brush-strokes, not the light of sight but the touch of touch. While the creamy square embodies the laws of the taken-for-granted, the half-and-half background serves the same function as the stone tablets onto which God once chiseled his ten commandments. And so it is that an icon is not *about* something, it *is* that something itself. The prohibition of prohibition, the taboo of the taboo.

More recently, the same limits of representation have been approached by Robert Ryman, not the least in his white oils on fibreglass with wax paper entitled *Versions I–XIV*. His own words about painting could easily be turned into mine about (post)modern Geography, an activity which is not content with being *about* something, but a mode of re-creation which at the same time strives *to be* that something (Storr 1993: 204): "In these paintings, the paint plane itself and the structure, being so thin, are large parts of the aesthetic. And then there is the very thin wax paper and the way that works with the light and the paint plane and the space of the environment that the painting is. All this has to do with the way the painting reacts with the wall plane. The wax paper, being very soft and impermanent, creates an aspect of fragility, the opposite of strength. Of course the paper can be replaced . . . There are also the nails that hold the paintings, and make them an essential part of the wall plane itself." An echo, not only of my own sentiments, but of Samuel Beckett's (1979: 352) Unnamable as well: "perhaps that's what I feel, an outside and an inside and me in the middle, perhaps that's what I am, the thing that divides the world in two, on the one side the outside, on the other the inside, that can be as thin as foil, I'm neither one side nor the other, I'm in the middle, I'm the partition, I've two surfaces and no thickness, perhaps that's what I feel, myself vibrating, I'm the tympanum, on the one hand the mind, on the other the world, I don't belong to either." The aesthetics of asceticism. Time–space compression of a different and higher order.

*

To the tune of these quotations from Robert Ryman and Samuel Beckett, the marcelverick Duchamp reenters the stage again. For now it is obvious that his *Large Glass – Mariée mise à nu par ses célibataires, même – The Bride Stripped Bare By Her Bachelors, Even –* is an elaborate study also of the tympanum of Plato's wall, the artist's dar(l)ing attempt to become one with the infra-thin tain of the projection plane. To appreciate the importance of that experiment, it should be noted explicitly that what first appears to be the front of the glass in reality

Figure 12.7 Marcel Duchamp, *Mariée mise à nu par ses célibataires, même*, 1915/23, reproduced by kind permission of the Philadelphia Museum of Art, Bequest of Katherine S. Dreier; © Succession Marcel Duchamp/ADAGP, Paris and DACS, London 2001

is its behind. More exactly, the original three-dimensional world of the lower half of the *Glass* is placed on the stage of the horizontal floor and then projected onto the vertical two-dimensional surface of the transparent screen that separates the stage from the auditorium, the spectators from the actors.

So strong is the impact that the Glider seems literally to break through the glass and come out on the other side.

Even more remarkable is the fact that the screen of the window pane is placed on the floor at a straight-up, 90-degree angle. If this were not the case, it would be extremely difficult to employ the standard techniques of linear perspective which Duchamp's delays were explicitly designed to illustrate. But even a genius of Duchamp's stature might occasionally be blind to the most obvious. Perhaps he simply failed to realize that the right angle is one of the most important human inventions ever made, an integral part of the taken-for-granted, the pivotal origo of Immanuel Kant's thesis about the necessary unity of consciousness, a principle doubly anchored in the politics of the group and the psychology of the individual. Without the fix-points of the perpendicular coordinate net almost everyone would be lost, craving for something to share and believe.

And via that reference to the geometrical inventions of Thales from Miletos, the exact location of Magritte's *Ceci* has just been determined. As always, the power-filled secret is hiding in the fraction line of the Saussurean Bar, in that razor-sharp boundary where the bodily senses of the Signifier become one with the cultural meanings of the signified. In the special case of the *Large Glass*, the merger occurs in the vanishing point of the Bachelors' perspective, itself positioned in the so-called Horizon, the line on which the Bride's Clothes are hung to dry. As explicated more fully elsewhere (Reichert 1996: 225–66), yet another self-portrait of *Rrose Sélavy*, Duchamp's alter ego, the homophone of the French expression "*Eros. C'est la vie.*" The similarities to El Greco's presence at the Count's funeral are striking. Dürer's reclining nude resurrected. '*Même*' sounds like '*m'aime*,' 'even' becomes 'love me.'

And so it is that all research is essentially autobiographical, the voyeur and the exhibitionist, the actor and the spectator, incestuously merged into one. But how do I practice a (post)modern geography as intense as Mark Rothko's paintings, so closely tied to the prohibited that many of them are named *Untitled*? Limit of limits, unbound desire muted by the lack of frames. Red on red, black on black, a brush so meticulously cleaned that it turns to silk. The touch untouchable, JHWH before JHWH. For Rothko himself, suicide was the only alternative. February 25, 1970.

*

Serious business. Laugh them away and the problems will return, every time as fresh as before. They certainly did for Marcel Duchamp, the man for whom the word 'amusing' was a favorite expression, the ironist who according to one source said that "there are no solutions because there are no problems," according to another that "there are no problems, there are only solutions." In the summarizing summary of his entire life – a construction on which he secretly worked from 1946 to 1966 – he went farther in his stripping bare of the taken-for-granted

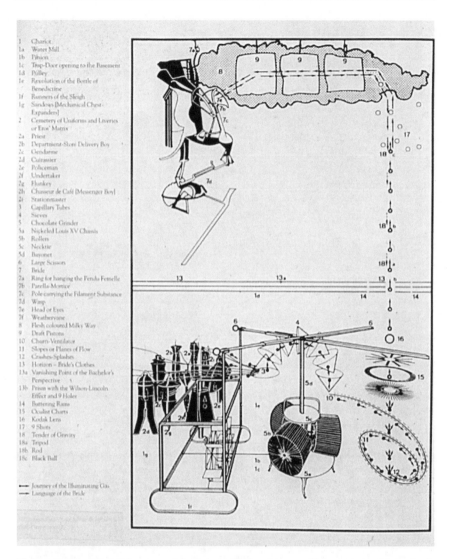

1   Chariot
1a Water Mill
1b Pinion
1c Trap-Door opening to the Basement
1d Pulley
1e Revolution of the Bottle of Benedictine
1f Runners of the Sleigh
1g Sandows [Mechanical Chest-Expanders]
2   Cemetery of Uniforms and Liveries or Eros' Matrix
2a Priest
2b Department-Store Delivery Boy
2c Gendarme
2d Cuirassier
2e Policeman
2f Undertaker
2g Flunkey
2h Chasseur de Café [Messenger Boy]
2i Stationmaster
3   Capillary Tubes
4   Sieves
5   Chocolate Grinder
5a Nickeled Louis XV Chassis
5b Rollers
5c Necktie
5d Bayonet
6   Large Scissors
7   Bride
7a Ring for hanging the Pendu Femelle
7b Parella-Mortice
7c Pole carrying the Filament Substance
7d Wasp
7e Head or Eyes
7f Weathervane
8   Flesh coloured Milky Way
9   Draft Pistons
10 Churn-Ventilator
11 Slopes or Planes of Flow
12 Crashes-Splashes
13 Horizon – Bride's Clothes
13a Vanishing Point of the Bachelor's Perspective
13b Prism with the Wilson-Lincoln Effect and 9 Holes
14 Battering Rams
15 Oculist Charts
16 Kodak Lens
17 9 Shots
18 Tender of Gravity
18a Tripod
18b Rod
18c Black Ball

— Journey of the Illuminating Gas
— Language of the Bride

Figure 12.8   Jean Suquet, *Itinerary of the Large Glass – Le miroir de la Mariée*, Paris, 1973

than ever before. Its French title: *Etant donnés: 1° la chute d'eau 2° le gaz d'éclairage.* In English: *Given: 1. The Waterfall 2. The Illuminating Gas.*

In this marvelous work, the uninitiated visitor encounters a wall with a walled-in door, a door which in effect is not a door at all but a

Figure 12.9 Marcel Duchamp, *Etant donnés: 1° la chute d'eau 2° le gaz d'éclairage*, 1946/66, exterior, reproduced by kind permission of the Philadelphia Museum of Art, gift of the Cassandra Foundation; © Succession Marcel Duchamp/ADAGP, Paris and DACS, London 2001

Figure 12.10 Marcel Duchamp, *Etant donnés: 1° la chute d'eau 2° le gaz d'éclairage*, 1946/66, interior, reproduced by kind permission of the Philadelphia Museum of Art, gift of the Cassandra Foundation; © Succession Marcel Duchamp/ADAGP, Paris and DACS, London 2001

window. Originally brought from Spain, it carries the exact proportions of the golden section. Since what looks like a door has no key, no handle, and no hinges, it cannot be opened. On closer inspection, however, one discovers two small holes drilled at eye level and placed 7 centimeters apart. Only the most courageous dare come close, dare to put their eyes to the holes, dare to turn the closed door into an open window. But when I do – when I tie myself to this solitary observation point – then I discover a world I never saw before. In the foreground of the diorama a three-dimensional nude on a bed of dried leaves and twigs, in the background a two-dimensional landscape with an electrically driven waterfall. So strong is the erotic power that it lights the gas lamp – Bec Auer – in the nude's left hand. All floors connected, the Bride's desire finally satisfied.

And, as I am standing there in this self-chosen position of the *Ceci*, I suddenly realize that I am at/of/in a pre-position which nobody else can share. Totally alone. Faintly remembering how the Bride of the *Large Glass* could be stripped bare by her many bachelors at the same time, I now experience how the truly *Given* is for one inspector only, how some truths cannot be shared. While Duchamp in the former work transformed Plato's limestone wall into transparent glass, in the latter he made the projection screen one with the viewer's eyes. Looked at from the side, the *Large Glass* is reduced to a one-dimensional line, a cut of anything whatsoever. In contrast, the *Given* refuses all attempts at reduction.

Every teenager knows where to look for the vanishing point. Like the formal logician he also knows that even though tautologies are always true, they are never informative. Whenever the need arises, a urinal is nevertheless a urinal, and that is regardless of the angle in which it is turned and of the name by which it is baptized. Meaning, though, is a different matter, for meaning is an idea, never visible, always mindful.

And that is why the foundation of the *Given* is not shown but hidden. Thus, under the nude's bed, invisible to the Peeping Tom, there is a floor. Not just any floor, however, but the black and white squares of the linoleum first lifted from Duchamp's New York kitchen and then translated to its final resting place in the catacombs of the Philadelphia Museum of Art. The Frenchman's reverence to ancestors like Brunelleschi and Masaccio, Velázquez and Vermeer, a reminder that there can be no children without parents, no *always* without a *not yet*. All performed by a devoted chess player, an individual who rather slept than wasted his time, a man so obsessed by prime numbers that he became a prime himself. A choreographer of thought-and-action. A (post)modern geographer in disguise. And Peggy Guggenheim used both

Figure 12.11 Richard Baquié, *Sans titre. Etant donnés: 1° la chute d'eau 2° le gaz d'éclairage...*, 1991, Collection of the Museum of Contemporary Art of Lyon. © Blaise Adilon

him and Samuel Beckett as her occasional lovers. Much ado about
nothing.

*   *   *

The point is made, the door is given. Peer through the holes! And
what will then be revealed is that the term 'postmodern geographical
praxis' is nothing but a legitimation of its own (anti)establishment, a
sanitized version of the politically correct. Dirty linens washed in a
faulty washing machine™. Clothes-lines stretched across the streets.
Banderoles waving to the crowds.

Such is the lasting memory from the colorful performances in the
Auditorium Santa Margherita. Anglo-American carpet bombing versus
Italian precision. How clear it became that quantity and quality are not
the same, that the essence is located at the minimalist periphery not at
the maximalist center. The Writing on the Wall: 1° the pre-positions,
2° the rhetorical tropes.

The foregone conclusion: Telling the truth is easy. Being believed is
likewise. The challenge is to do both at the same time, to be one with
the tympanum, not to follow the swings of the pendulum but to be
lodged in the fulcrum on which the pendulum is hung. Fix me a point
and I shall move the world.

(Post)modern Geographical Praxis deconstructed. No limit sacred, no
member untouched. The imperative is in the interface of the I and the
Thou. It was Ludwig Wittgenstein (*Tractatus*, 6.421) who once wrote
that "ethics and aesthetics are one and the same."

He was right. Or, to be internally consistent, so I believe.

REFERENCES

Beckett, S. (1979). *The Beckett Trilogy: Molloy, Malone Dies, The Unnamable.*
London: Picador.
Beckett, S. et al. (1929). *Our Exagmination Round His Factification
for Incamination of Work in Progress.* Paris: Shakespeare and
Company.
Breslin, J. E. B. (1993). *Mark Rothko: A Biography.* Chicago: Chicago University Press.
Cabanne, P. (1977). *Dialog med Duchamp.* tolkn. Jan Östergren Lund: Bo
Cavefors.
Cabanne, P. (1997). *Duchamp & Co.* Paris: Éditions Terrail.
Camfield, W. A. (1989). *Marcel Duchamp: Fountain.* Houston: Houston Fine
Arts Press.
Cézanne, P. (1984). *Letters,* ed. J. Rewald. New York: Hacker Art Books.

Clair, J. (1975). *Marcel Duchamp ou le grand fictif: Essai de mythanalyse du grand verre*. Paris: Éditions Galilée.

Clair, J. (1978). Duchamp and the Classical Perspectivists. *Artforum* 16: 40–9.

Cough-Cooper, J. and J. Caumont (1993). *Marcel Duchamp – Ephemerides on and About Marcel Duchamp and Rrose Sélavy, 1887–1968*. Exhibition Catalog, Palazzo Grassi, Venezia Milan: Bompiani.

Crone, R. and D. Moos (1991). *Kazimir Malevich: The Climax of Disclosure*. London: Reaktion Books.

Damisch, H. (1994). *The Origin of Perspective*. Cambridge, Mass.: MIT Press.

Deleuze, G. (1993). *The Fold: Leibniz and the Baroque*. Minneapolis: University of Minnesota Press.

Deleuze, G. and F. Guattari (1987). *A Thousand Plateaus: Capitalism and Schizophrenia*. Minneapolis: University of Minnesota Press.

Derrida, J. (1987). *The Truth in Painting*. Chicago: University of Chicago Press.

Doel, M. A. (1999). *Poststructuralist Geographies: The Diabolical Art of Spatial Science*. Edinburgh: Edinburgh University Press.

Duchamp, M. (1968). I Propose to Strain the Laws of Physics. Interview by Francis Roberts. *Arts News*, Dec.: 67.

de Duve, T. (1991). *Pictorial Nominalism: On Marcel Duchamp's Passage from Painting to the Readymade*. Minneapolis: University of Minnesota Press.

de Duve, T. (1996). *Kant after Duchamp*. Cambridge, Mass.: MIT Press.

Elkins, J. (1994). *The Poetics of Perspective*. Ithaca, NY: Cornell University Press.

Farinelli, F. (1992). *I segni del mondo. Immagine cartografica e discorso geografico in età moderna*. Florence: Nuova Italia Editrice.

Farinelli, F., G. Olsson, and D. Reichert, eds. (1993). *Limits of Representation*. Munich: Accedo.

Foucault, M. (1983). *This is not a Pipe*. Berkeley: University of California Press.

Gasché, R. (1986). *The Tain of the Mirror: Derrida and the Philosophy of Reflection*. Cambridge, Mass.: Harvard University Press.

Gregory, D. (1994). *Geographical Imaginations*. Oxford: Blackwell.

Guggenheim, P. (1979). *Out of this Century: Confessions of an Art Addict*. London: André Deutch.

d'Harnoncourt, A. and K. McShine, eds. (1973). *Marcel Duchamp*. Exhibition Catalog, Philadelphia Museum of Art. New York: Museum of Modern Art.

Harvey, D. (1989). *The Condition of Postmodernity: An Enquiry into the Origins of Cultural Change*. Oxford: Blackwell.

Kandinsky, W. (1977). *Concerning the Spiritual in Art*. New York: Dover.

Kandinsky, W. (1971). *Point and Line to Plane*. New York: Dover.

Kant, I. (1989). *The Critique of Judgement*. Oxford: Clarendon Press.

Koerner, J. L. (1993). *The Moment of Self-Portraiture in German Renaissance Art*. Chicago: University of Chicago Press.

Lyotard, J.-F. (1977). *Les trans formateurs Duchamp.* Paris: Éditions Galilée.

Linde, U. (1986). *Marcel Duchamp.* Stockholm: Rabén and Sjögren.

Malevich, K. (1968). *Essays on Art.* Copenhagen: Borgen.

Mallarmé, S. (1965). *Mallarmé.* Harmondsworth: Penguin.

Meri, V. (1964). *Kvinna i spegeln.* Helsinki: Söderström.

Meuris, J. (1997). *René Magritte, 1898–1967.* Cologne: Taschen.

Nehamas, A. (1985). *Nietzsche: Life as Literature.* Cambridge, Mass.: Harvard University Press.

Novotny, F. (1938). *Cézanne und das Ende der wissenschaftlichen Perspektive.* Vienna: Schroll.

Olsson, G. (1980). *Birds in Egg/Eggs in Bird.* London: Pion.

Olsson, G. (1991). *Lines of Power/Limits of Language.* Minneapolis: University of Minnesota Press.

Olsson, G. (1994). Heretic Cartography. *Ecumene* 1: 215–34.

Olsson, G. (1997). Misión imposible. *Anales de Geografía de la Universidad Complutense* 17: 39–51.

Olsson, G. (1998). Towards a Critique of Cartographical Reason. *Ethics, Place and Environment* 1: 145–55.

Olsson, G. (2000). Skattkammarön. In *Kulturens plats/Maktens rum,* eds. M. Gren et al. Stehag: Symposion.

Panofsky, E. (1991). *Perspective as Symbolic Form.* New York: Zone Books.

Panofsky, E. (1955). *The Life and Art of Albrecht Dürer.* Princeton: Princeton University Press.

Paz, O. (1978). *Marcel Duchamp: Appearance Stripped Bare.* New York: Viking Press.

Plato (1968). *Republic.* New York: Basic Books.

Pred, A. (2000). *Even in Sweden: Racisms, Racialized Spaces and the Popular Geographical Imagination.* Berkeley: University of California Press.

Reichert, D., ed. (1996). *Räumliches Denken.* Zürich: vdf Hochschulverlag an der ETH Zürich.

Rotman, B. (1987). *Signifying Nothing: The Semiotics of Zero.* London: Macmillan.

Rousel, R. (1967). *Impressions of Africa.* London: Calder and Boyars.

Schapiro, M. (1988). *Paul Cézanne.* New York: Harry N. Abrams.

Scherer, V. (1908). *Dürer: Des Meisters Gemälde, Kupferstiche und Holzschnitte.* Stuttgart: Deutsche Verlags-Anstalt.

Seigel, J. (1995). *The Private Worlds of Marcel Duchamp: Desire, Liberation and the Self in Modern Culture.* Berkeley: University of California Press.

Serres, M. (1993). *Les origines de la géométrie: Tiers livre des fondations.* Paris: Éditions Flammarion.

Soja, E.W. (1989). *Postmodern Geographies: The Reassertion of Space in Critical Social Theory.* London: Verso.

Storr, R. (1993). *Robert Ryman.* Exhibition Catalog. London: Tate Gallery.

Suquet, J. (1974). *Miroir de la Mariée.* Paris: Éditions Flammarion.

Suquet, J. (1991). *Le Grand Verre rêvé*. Paris: Éditions Flammarion.
Taylor, M. C. (1997). *Hiding*. Chicago: University of Chicago Press.
Weiss, J. et al. (1998). *Mark Rothko*. Exhibition Catalog, National Gallery of Art, Washington. New Haven: Yale University Press.
Wittgenstein, L. (1961). *Tractatus Logico-Philosophicus*. London: Routledge and Kegan Paul.

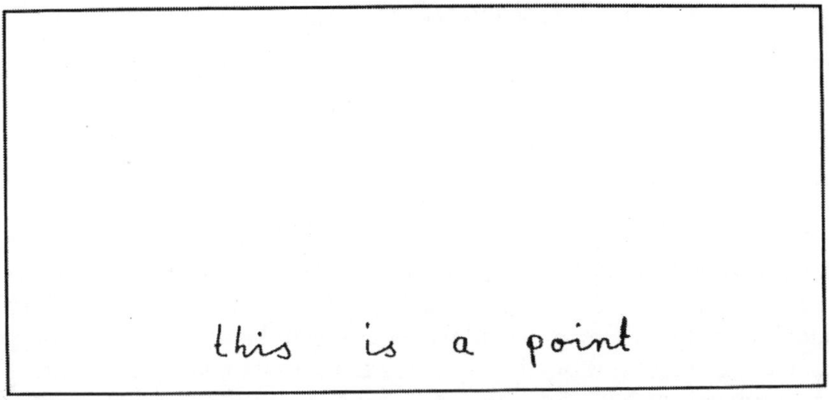

Figure 12.12 Gunnar Olsson, *This is a point*, 1987, reproduced from Olsson 1991: 65

Afterword

Edward W. Soja

What can be learned from a retrospective glance back at the preceding twelve essays? At the very least, such a reflective effort makes it abundantly clear that there are – and will continue to be – multiple and often conflicting definitions of all three of the terms that make up postmodern geographical praxis. Such a conclusion is neither surprising nor unhealthy, for few would disagree that a diversity of viewpoints, and indeed some passionate polemical contention, can be highly stimulating. Claudio Minca's excellent introduction takes this optimistic position, celebrating the confrontational diversity emerging from the dozen responses and reactions to what is now widely described as the "postmodern turn" in geography.

That there exists so much disagreement and contention in the essays is nevertheless surprising, given that there was reason to expect much greater underlying unity from the twelve authors who originally came together to discuss postmodern geographical praxis in Venice on a few hot days in June, 1999. All those presenting papers at that conference were identifiable as critical human geographers, and many had participated centrally in defining this identity over the past two decades. At least half did not actually teach in departments of geography, an indication of their strong cross-disciplinary linkages, interests, and orientations. In terms of personal knowledge, the six North American geographers overlapped significantly with the European half dozen. What then produced such a diversity of perspectives and approaches?

In part, such diversity simply reflects the eclecticism and plurality of viewpoints that is typically associated with postmodernism and postmodernity. But the essays here are not merely eclectic and diverse, they expose of series of deep cleavages that have fractured the praxis of postmodern geography into what appear to be incompatible, if not conflicting, approaches. Never before has it been clearer that the challenge of postmodern geographical praxis serves both to unite and divide, to foster mutual understanding as well as uncompromising conflict and disagreement. Rather than simply welcoming the diversity or trying to find some basis for consensus, I will use this Afterword to reflect upon some of the major points of contention that differentiate the various approaches to postmodern geographical praxis presented in this volume. These divergences and differences are likely to be more revealing than any attempt to find a unified voice.

Continental Divides

One of the least contentious divisions evident in this collection of essays is brought out in Gunnar Olsson's ever churning washing machine. His observations agitated the final moments of the Venice conference and now stand, in revised form, as the concluding essay of this volume. Building on the pre-positional geographies condensed in the two words 'of' and 'in,' Olsson (with a nod to Samuel Beckett's remark about James Joyce) draws a contrast between those driven to write about (of) postmodern geographical praxis and those whose aim is to produce a text that embodies, that demonstrably *is* (in) such a praxis. With perhaps too broad a brush stroke, Olsson associates this stylistic and epistemological division between 'talkers' and 'doers' with Anglo-American versus European sensibilities.

Sitting squarely in the Joycean camp, Olsson takes this methodological (and continental) contrast to its apparent extreme in his own essay, which departs from any simple summing up to stake out a distinctive path of its own. Following closely the mainly European section on *Mappings*, Olsson takes us all into a self-referential borderworld between semiotics and politics to "cultivate the fragment," to deconstruct (the possibility of) postmodern geographical praxis sufficiently to conclude that it is "far too radical for the majority of self-appointed radicals" who treat it as little more than a "sanitized version of the politically correct." This is a direct reference to what he describes as the "colorful performances" in Venice of certain Anglo-American geographers "carpet bombing" one another to get on to the definitive high ground.

But these North American allusions are merely sidebars to Olsson's indulgent excursion through a field-full of ironic ambiguities selectively seen through the artfulness of Duchamp and Magritte, the masters of cultured gamesmanship. Whether this journey takes us into postmodern geographical praxis or leads us further away from it is unclear, purposefully ambiguous. Nor does it seem to matter to Olsson. Direct references to the postmodern are few and strategically obscured. Note, in particular, his insistence on parenthetically floating the 'post' as (post)modern, suggesting that one could take it or leave it. Perhaps the parentheses are meant to show that the modern and the postmodern co-exist, one in the other and not in categorical opposition, as so many critics of postmodernism presuppose? But they can also mean, with typical postmodern irony, that it doesn't matter very much what you call it, and, in any case, there is no reason we should spend much time talking and writing about it.

Of more concern to Olsson is an active commitment to discursive experimentation, rhetorical twists and turns that disrupt the taken-for-grantedness of the printed text, that rattle established – and presumably modern – conventions. And with delightful ambiguity he continues to practice what he preaches, as he has for decades, always finding new ways to make you wonder at his uncanny ability to make you feel hopelessly out of the (his) picture. But one must not be side-tracked by Olsson's playful text, for there is more – and less – to the apparent division between Anglo-American talkers and continental European doers than initially meets the eye.

No one other than Olsson engages in any significant textual experimentation. What most distinguishes Olsson's in-group from his of-group is, first, that the doers take for granted the existence of a critical (post)modern (or at least counter-hypermodern) geographical praxis; and second, that they avoid talking very much about what it is, and instead proceed to apply its insights, methods, and fundamental epistemological critique – whatever best can be taken from this critical praxis – to broad problems of contemporary geographical understanding and analysis. Franco Farinelli comes closest to Olsson's viewpoint and style, not surprisingly as the two have worked together for many years. He carries forward under the (post)modern banner their joint project aimed at a neo-NeoKantian post-existentialist Critique of Cartographical (cum Geographical) Reason that brings the crisis-riddled discipline of Modern Geography closer, in Farinelli's view at least, to the uncertain atmo(sphere) of quantum physics. Ambiguous precision reigns in Farinelli's quantumized postmodern world of mythic uncertainties, rooted nonlocality, unmappable territories, a world where the real and the imagined seamlessly intertwine. Classic geography and cartographic logic are dissolved in the spaces between the territory and the map, with a seeming nod to the hyperrealities of Umberto Eco, Farinelli's close colleague at the University of Bologna.

Vincenzo Guarrasi follows a similar (post)modern reading of space by focusing on the real and imagined, material and metaphorical landscapes, inviting us to "suspend belief" in the cartographic (and geographic) logic of modernity and open up new and subversive "ways of seeing" human spatiality. Key to this new way of seeing for Guarrasi is Foucault's concept of heterotopia, which pops up with radically different definitions and purposes in several other essays as one of the most indicative icons of postmodern geographical praxis. Here its use is closer than these other references, I think, to Foucault's intended meaning, which refers not so much to a particular kind of space but to a particular way of looking at and interpreting human spatiality more generally (Soja 1996). Hence Foucault's emphasis on the more encompassing notion of *heterotopology*, an alternative and decidedly different (postmodern?) way of looking at and interpreting space and geography, landscapes and maps, sites of struggle and recovery, and, for Guarrasi, even that obsessive presence in contemporary geography, Geographical Information Systems (GIS).

The essays by Olsson, Farinelli, and Guarrasi can be tied together to define a consistent and distinctive positioning on postmodern geographical praxis, a point of view that arises from an assumption that 'Modern Geography' is (necessarily) in crisis. It may be too early (and probably not very useful) to announce the triumph of postmodernism, and hence the caution with which the term is used. Although it is not stated as such, there are enough indications to suggest what critical human geographers should be doing is *practicing heteropology*, another way of seeing-doing-writing that differs significantly from well-established modern geographical practices, whether they be positivist, humanist, Marxist, phenomenological, or otherwise. From this viewpoint, (post)modern geographical praxis is not a simple and unproblematic offshoot from the traditions of modern geographical thought, a bunch of new ideas and concepts that can be usefully reabsorbed into established frameworks and schools of geographic thought. It is, or can be made to be, a deeply subversive and transformative critique that strikes at the heart and soul of Modern Geography: the way in which we perceive, conceptualize, experience, and map space in its multiple forms, as body, site, place, locality, landscape, scale, neighborhood, territory, city, region, world . . .

The Italian version of this heterotopology, the search to escape from the closed spaces framed by the logic of modernity, is built upon and extended in the chapter by Claudio Minca. Venice-Italy meeting Venice-California (with the latest Venice simulation in Las Vegas also in mind) forms the backdrop to Minca's "new exploration" of *des espaces autres*, in this case the exhibitional spaces of mystification that set up the postmodern and postcolonial world as simultaneously and selectively real and imagined, enclosed in a modernist metaphysics of representation. Also stretching beyond such a modernist metaphysics, as well as

shifting from a focus on *Mappings* to one on *Scales* and *Cities*, Giuseppe
Dematteis completes the distinctively Italian tour of post-Modern Geography by alighting on a new urbanism, which he describes as fragmented, polarized, networked on a global scale, uprooted, incoherent,
complex, dissolved into images. Dematteis explicitly politicizes his
insights more than the other continental essays, arguing that we need a
new geographical praxis, however it might be described, to contend with
the territorial injustices inherent in the postmodern city.

There are many links here to the Anglo-American essays, especially
in exploring, from radically different perspectives, the impact of what
is alternatively called the new urbanism, postmodern urbanism, or the
restructuring of the modern metropolis. Indeed, if there is any common
theme that appears in nearly all the essays and cuts across Olsson's mid-
Atlantic divide, it relates to the dramatic urban restructuring and related
globalization processes reshaping geographies all over the world during
the past thirty years. The Anglo-American geographers, however, are
much more conscious of the fundamental differences that exist among
them in their interpretation of these restructuring processes, and are
much more inclined to write about these differences explicitly and
directly. In contrast, the European contingent tends to present a much
more united front, practicing their postmodernism without feeling it
necessary to dwell on whatever fundamental differences may exist
among them. But, in the end, the alleged continental breach between
talkers and doers lands up being less well defined and unbridgeable than
Olsson seems to assume.

The continental contrasts noted by Olsson nonetheless serve a useful
purpose, in that they make it easier to recognize the special qualities of
the essays written by Italian geographers. It is ironic, for example, that
some of the most insightful and creative contemporary critiques of
Modern Geography are coming from one of the most devastated fields
of Modern Geography in all of Europe. It becomes even more interesting to discover that so many of the best critical human geographers
remaining in Italy today are able to come together so comfortably and
creatively under the rubric of postmodern geographical praxis. In any
case, hearing the critical intellectual voices of the Italian geographers
was a highlight of the Venice conference for me, and these voices now
represent a crucial contribution to the present volume.

Bicoastal Division

Implicit in all but a few of the twelve essays, as either an assumption
or a challenge, is the argument that postmodern geographical praxis

involves not just an acute sensitivity to the distinctiveness of the present but also a belief that what is new and different here and now requires a significant restructuring of established, modernist ways of under-standing and acting in the contemporary world. As a critique of Modern Geography, this critical postmodern approach has tended to operate in three interrelated ways: as a deconstructive *method* or language of textual and visual representation; as an interpretation of postmodernity as a distinctive empirical and geographical *condition* of the contempo-rary world; and, perhaps most demanding and difficult, as a compre-hensive *epistemological critique* that calls for a radical rethinking and revisioning of all established forms of geographical practice.

All of these assumptions and approaches, with varying emphasis to be sure, are interwoven in the European cluster of essays, including the open-minded "performative" exploration of "millennial geographies" by Denis Cosgrove and Luciana Martins. Indeed, they are virtually taken for granted. They are also explicitly present in my essay explor-ing the "postmetropolis" and in Michael Dear's opening chapter chro-nicling the heady debates on modernism and postmodernism in Geography and Planning over the past twenty years. But in these first two chapters, the critical assumptions and methods are more explicitly and discursively argued and advocated rather than taken as given, sup-porting Olsson's continental contrast between talkers and doers. At the same time, it is important to recognize that this contrast is not so much the product of a difference in intellectual cultures, as implied by Olsson, but of the fact that postmodernism has had a much deeper impact and has, at the same time, been much more highly contested in Anglo-American geography than it has on the European continent.

Not only have there arisen within Anglo-American geography rather virulent forms of *anti-postmodernism*, some of the strongest attacks have come from scholars who many other geographers have persisted in associating, in a positive way, with postmodernism, the most notable being David Harvey and, as Dear notes in his chapter, Alan Pred, whose preference for the term hypermodernism versus postmodernism is at least partially shared by some of the continental European geographers represented here. Great ambiguity and confusion thus continue to exist along the outer boundary of advocative postmodern geographical praxis, creating formidable barriers between geographers who otherwise share much the same intellectual and political perspective. These con-ditions were aggravated further in Venice by the pre-conference avail-ability of Michael Dear's ecumenical but highly personalized chronicling of the insistent rise of academic postmodernism in Anglo-American geography, and then his non-appearance at the conference to hear his critics respond (because of a broken ankle). This created certain hypersensitivities that would boil over in the acrimonious debate that Olsson referred to in his essay.

Growing out of these tensions, the Anglo-American essays in this book can be easily split into a more-than-stylistic opposition between those accepting the premises of postmodern geographical praxis and trying to learn from them (Dear, Soja, Flusty, and probably also Cosgrove and Martins) as opposed to those who see postmodern geographical praxis as unavoidably associated with the exploitative and oppressive practices of globalized capitalism (Mitchell, Smith, and Katz, although the latter two much more so in their original presentations than in their present chapters). While the first group talk about and actively practice postmodern geography, the second are almost irresistibly compelled to attack what they see and hear as either solipsism (in the talking) or duplicitous if unintended attachment to the darkest forces of contemporary capitalism (in the practice).

What specifically triggered the acrimony in Venice was the issue of simplicity versus complexity. Faced with a barrage of papers dwelling on the challenging complexities of contemporary postmodernity and the claim by Dear that we were all, in one way or another, postmodern geographers, the New York area contingent attacked (Olsson called it carpet bombing) their Southern California opponents on the grounds that all this talk of complexity was burying the basic political simplicity of yes versus no, acting in favor of the oppressed classes /genders/races or against them. The implications were clear enough, and they extended well beyond the simplicity/complexity battle. Practicing postmodern geography, in the minds of the East Coast geographers, was a divisive and potentially dangerous political diversion, a submission to the oppressive forces of globalization and neoliberalism, which to this more orthodox group of Marxist geographers were the very definition, the essence, of postmodern geographical praxis in the material, empirical world. All this talk about epistemological critiques and the need to revise established geographical practices, especially if turned against Marxist Geography, was little more than useless wordplay or worse, political self-delusion.

The Italian geographers were not the intended targets for the attacks on the complexifiers, even though they, more explicitly than others both at the conference and in the present collection of essays, specifically recognize and write about the challenging complexities of the contemporary world. But as hosts, they were deeply embarrassed at the blunt intensity of some of the personal attacks, and tried hard to find some balance (and respect) among the two opposing sides, an effort that is reflected in Claudio Minca's *Prelude*. It is clear, however, that the confrontation is not entirely about simplicity versus complexity, but reflects much deeper theoretical, practical, and political divisions. Just below the surface, for example, is a longstanding contrast in intellectual cultures between critical geographers whose work focuses on New York and other East Coast representations of what is lost often called post-industrialism, and those that are centered in Los Angeles, where such terms as postmodern and postfordist are more easily accepted. The homogeneous intellectual culture that is projected on North America by Olsson turns out to contain even greater contrasts and oppositions than were assumed to exist between European and North American critical geographical traditions.

The conflicts that came to the surface in Venice persist, albeit with some significant softening, in the present set of essays. The clash of viewpoints revolves around several pronounced contrasts. In what for argument's sake can be called the New York view, there is a tendency to see too much of the writing and reading of geographies by those who seem sympathetic to postmodernism and focused on Los Angeles as a reduction to pure style, to language games and esoteric textual play that have the depoliticizing effect of drawing attention away from the serious problems embedded in the contemporary capitalist world. Similarly, aggressive epistemological critiques, especially when applied to established forms of Marxist geographical analysis, and emphasis on the immaterial (images, simulacra, hyperreality) also work to divert, depoliticize, and overly complicate what is essentially simple and straightforward. If the condition of postmodernity is, as David Harvey (1989) put it, nothing more than capitalism speeded up, then to think otherwise, to even imagine the hope of any liberatory potential, makes one an unwitting accomplice to its egregious survival and expansion.

In the view from Los Angeles, the New York perspective is a misleading and misinformed reduction of postmodernism only to the material conditions and practices associated with global capitalism. Such a narrowing not only makes it seem that everyone espousing postmodernism in any form is inherently in cahoots with global capital and its henchmen, it effectively side-steps the powerful epistemological and ontological critiques of Modern Geography in all its forms, but especially with regard to Marxist geographical analysis. This anti-postmodernism protects a fundamentally flawed form of Marxist analysis from significant self-questioning, and leads too often to grossly oversimplified applications of rigid Marxist concepts and approaches, such as explaining every evil in sight simply as the product of a seemingly endless crisis of social production and reproduction. There is no doubt that globalization, urban restructuring, and predominant forms of (neoliberal) postmodern political practice have aggravated social inequalities, increased homelessness, and led to intensified forms of oppression and exploitation all over the world, but, yes, the processes of change are too complex to be understood simply by drawing only on the volumes of *Capital* and a still only partially geographical historical materialism.

This bicoastal split is most boldly expressed in the stark contrasts between the essays by Michael Dear and Don Mitchell. Dear's discussion of the postmodern turn seems resolutely nonpolitical. There is no mention of the contemporary politics of postmodernity, for everything is made to hinge around intradisciplinary debates (who said what and when) within the confines of Geography and Planning. The postmodern turn is presented from such an introverted perspective that it can too easily be seen as depoliticizing rhetoric that endlessly talks about postmodern geography without ever actually practicing it or appreciating its wider implications.

The essay by Don Mitchell flips wildly to the other extreme, depicting all the Los Angeles (as well as other) postmodernists, at their best, as faddish and slightly effete stylists. At their worst, they become compliantly lumped together with the dominant capitalist practitioners of postmodernism, with all references to such ephemera as heterotopias and thirdspaces serving only to mask the most oppressive conditions of postmodernity. Dear and I, despite what we have written elsewhere on the subject, are blithely transformed into "handmaidens" to the "brutal postmodern geographical practice" that aims at eliminating not homelessness but the lives of homeless people themselves. Under these erratically imagined conditions of postmodernity, Mitchell feels free to engage in what can be best described as textual terrorism, since no holds need be barred (including at times the normal rules of respectable academic discourse) in attacking all those misled evildoers who dare to dwell in an alternative world, in other spaces and places, in different geographies from his own. So absorbed is Mitchell with blaming others that he offers almost no insight at all into the contemporary tragedy of homelessness and the housing crisis, and how they might be ameliorated.

That this contested terrain expressed itself in East versus West, New York versus Los Angeles clusters may in part be a reflection of contrasting attitudes to continuity versus change, and perhaps also to time and history versus space and geography, on the two coasts. Older and deeper traditions and a more vivid sense of the past encourage a greater appreciation for historical continuities along the northeast coast; while the relative newness of large-scale settlement, greater locational and environmental diversity, and possibly also recent economic expansion has produced a greater sensitivity to change, to departures from longstanding trends in southern California. Even bringing up such geographical explanations may stamp me with an LA label and I do not want to suggest some bicoastal determinism, but in the half dozen chapters by North American writers, the New York–Los Angeles split is too obvious to ignore and too easy to explain by clashing personalities and politics alone.

The three Los Angeles papers (Dear, Soja, Flusty) emphasize *what is new and different* in the contemporary world and are infused with a positive commitment to postmodern geographical praxis. All three have written, in particular, about postmodern urbanism and would have no trouble agreeing, to use Claudio Minca's introductory words, that we are "now confronted with revolutionary geographies . . . that are radically different from those of modernity – and thus necessitate new analytical tools, new understandings capable of grasping their dynamics." In an interesting way, with their emphasis on what is new and different, indeed heterotopological, about the contemporary world, these three papers share much more with the European half dozen (with Cosgrove, now Humboldt Professor of Geography at UCLA, forming an Anglo-American bridge) than the East Coast three.

The latter (Mitchell, Smith, Katz) are assertively Marxist (ortho-topological?) and as such are particularly effective in analyzing the continuities of capitalism, *what is fundamentally the same* today in an age of globalisation, neoliberalism, flexible accumulation, and what might be described as the 'rush to the post.' In this view, there are certainly no new 'revolutionary geographies' out there, and that which is being promoted as new becomes the primary target for the three essays, ranging from Mayor Giuliani's programs and projects to globalization, the new urbanism, the cultural turn, and postmodern geographical praxis itself. What others see as new, complex, and challenging in the contemporary world are seen here as essentially known quantities dressed up in deceptive ideological imagery. Thus there is no need to seek new analytical tools and different modes of understanding, for the old ones (or at least a historical materialism that is also geographical) are good enough to deal with contemporary reality, just as they have been in the past. All that is needed is effective demystification to expose the hidden worlds beneath the surface.

Among the New York essays, the only significant bow to the other coast or, for that matter, the other side of the Atlantic as well, is Cindi Katz's use of Foucault's concept of heterotopia, which contrasts sharply with her East Coast colleagues' use of the same term. While her main argument continues to be that global capitalism, as an insidious form of postmodern geographical praxis, is deeply restructuring social relations of production and reproduction, altering our connections to particular places and spaces, and hiding it all from us under veils of postmodern imagery, Katz at the same time recognizes that these hidden violences and oppressions can be revealed and brought to the surface through the concept of heterotopia and a critical version of postmodern geographical praxis, something which Mitchell and Smith, at least in these essays, seem to believe cannot exist. But what connects her back into the New York fold is another manifestation of the simplicity/complexity debate, a sense of nearly complete confidence that one knows what will be revealed or unhidden when the layers of mystification are removed.

A similar self-confidence shapes Smith's essay on rescaling politics and his take on the new urbanism. Although a refreshing advance on his original presentation, nowhere in the essay is there even a hint that a critical and radical postmodernism might be possible, much less that it can be insightfully applied, even if only to help demystify the contemporary world as Katz suggests in her discussion of the concept of heterotopia. It may appear that the all-embracing concept of rescaling he proposes is an example of his openness to new approaches and analytical tools in responding to the conditions of postmodernity, but its interpretation and presentation, as well as the overconfident tone of the essay as a whole, derives almost entirely from an anti-postmodernist Marxian political economy mildly spatialized and protected from such insidious diversions as the cultural turn and the critique of historicism, both of which receive snide asides.

What then of the argument raised by Claudio Minca in his *Prelude* stressing the need for diverse voices – including the challenges posed by the New York contingent to more prevalent forms of postmodern geographical praxis? All I can say in response is that I wish they had done the job better, for there is much to criticize in the contemporary practices of postmodern and (post)modern geographers.

False Unities

Running through nearly all the essays, giving a deceptive surface appearance of connectedness and unity, are two related features of the contemporary world that have become virtually equated with postmodern geographical praxis. The first can be generally described as postmodern urbanism, based on the recognition that most cities have been experiencing a dramatic restructuring over the past thirty years and that this restructuring has produced reorganized geographies that demand our theoretical and practical attention. The essays by Dematteis, Soja, and Smith deal most specifically and directly with the challenge of making theoretical and practical sense of these new urban geographies, but a similar focus can be found in the chapters by Katz, Minca, Cosgrove and Martins, Mitchell, Flusty, and Dear. What this widespread interest suggests is that the subfield of urban geography has received more attention than any other in the wake of Geography's postmodern turn. The approaches taken to interpreting this postmodern urbanism are exceedingly diverse, especially in this volume, but it would not be inaccurate to assume that the primary empirical focus of postmodern geographical praxis has been the city and the changes that have been taking place in urban life, especially in large 'world cities,' over the past three decades.

The second feature characterizing nearly all the essays has been some form of conflation between postmodernity (or, as a process, postmodernization) and globalization (or, one might say, the condition of globality). There are a number of contemporary scholars who now argue that the discourse on globalization has replaced the longstanding debates about modernity and postmodernity on the agenda of critical social theory, and to some degree such a substitution occurs in many of the essays included here. Moreover, the voracious globalization process is widely assumed to be the primary force behind the development of postmodern urbanism. This conflation of postmodernism and globalization can be found in many of the essays by Italian geographers and is rather oddly shared by Steven Flusty and Neil Smith in their markedly different essays.

For Flusty, improving our understanding of the everyday and individualized effects of globalization is assumed to be, in itself, a form of postmodern geographical praxis. Neither globalization nor postmodern urbanism is explicitly described as unequivocally good or bad, but rather as existing and not to be ignored. For Smith, globalization and the restructuring of geographical scale becomes the prime example of capitalism's postmodern geographical praxis, thus setting such praxis up as a trap for radical geographers and geography. In this view, globalization becomes inherently bad and, therefore, to be against globalization (for example, in the recent struggles against the WTO in Seattle) is also to be against postmodernism. In this way, the two essays use a shared assumption about globalization to reproduce the most confrontational opposition dividing the contributors to this volume, further confusing the debate on postmodern geographical praxis.

In my own essay (see also Soja 2000), I describe globalization as one of a series of processes reshaping cities and regions today, and as a major (but not the only) force contributing to the development of postmodern urbanism. But rather than extend this argument any further, it is sufficient to merely state here my conviction that the conflation of postmodernism with globalization is a mistake and should be avoided for

many different reasons, not the least of which is the political and theoretical confusion it causes. The vigorous advancement of a critically
informed and progressive postmodern geographical praxis must resist
such simplistic conflations and contest the definitions imposed on it by
anti-postmodernists. Perhaps the best – and last – bit of advice that can
be given to the readers of this book is to not accept anyone else's definition of postmodern geographical praxis.

REFERENCES

Harvey, D. (1989). *The Condition of Postmodernity*. Oxford: Blackwell.
Soja, E. W. (1996). *Thirdspace: Journeys to Los Angeles and Other Real-and-
Imagined Places*. Oxford: Blackwell.
Soja, E. W. (2000). *Postmetropolis: Critical Studies of Cities and Regions*.
Oxford: Blackwell.

Index